T0172790

COMPUTATIONAL
BIOLOGY
A STATISTICAL MECHANICS
PERSPECTIVE

CHAPMAN & HALL/CRC
Mathematical and Computational Biology Series

Aims and scope:

This series aims to capture new developments and summarize what is known over the whole spectrum of mathematical and computational biology and medicine. It seeks to encourage the integration of mathematical, statistical and computational methods into biology by publishing a broad range of textbooks, reference works and handbooks. The titles included in the series are meant to appeal to students, researchers and professionals in the mathematical, statistical and computational sciences, fundamental biology and bioengineering, as well as interdisciplinary researchers involved in the field. The inclusion of concrete examples and applications, and programming techniques and examples, is highly encouraged.

Series Editors

Alison M. Etheridge
Department of Statistics
University of Oxford

Louis J. Gross
Department of Ecology and Evolutionary Biology
University of Tennessee

Suzanne Lenhart
Department of Mathematics
University of Tennessee

Philip K. Maini
Mathematical Institute
University of Oxford

Shoba Ranganathan
Research Institute of Biotechnology
Macquarie University

Hershel M. Safer
Weizmann Institute of Science
Bioinformatics & Bio Computing

Eberhard O. Voit
The Wallace H. Couter Department of Biomedical Engineering
Georgia Tech and Emory University

Proposals for the series should be submitted to one of the series editors above or directly to:
CRC Press, Taylor & Francis Group
24-25 Blades Court
Deodar Road
London SW15 2NU
UK

Published Titles

Cancer Modelling and Simulation
Luigi Preziosi

Computational Biology: A Statistical Mechanics Perspective
Ralf Blossey

Computational Neuroscience: A Comprehensive Approach
Jianfeng Feng

Data Analysis Tools for DNA Microarrays
Sorin Draghici

Differential Equations and Mathematical Biology
D.S. Jones and B.D. Sleeman

Exactly Solvable Models of Biological Invasion
Sergei V. Petrovskii and Lian-Bai Li

An Introduction to Systems Biology: Design Principles of Biological Circuits
Uri Alon

Knowledge Discovery in Proteomics
Igor Jurisica and Dennis Wigle

Modeling and Simulation of Capsules and Biological Cells
C. Pozrikidis

Normal Mode Analysis: Theory and Applications to Biological and Chemical Systems
Qiang Cui and Ivet Bahar

Stochastic Modelling for Systems Biology
Darren J. Wilkinson

The Ten Most Wanted Solutions in Protein Bioinformatics
Anna Tramontano

Chapman & Hall/CRC Mathematical and Computational Biology Series

COMPUTATIONAL BIOLOGY
A STATISTICAL MECHANICS PERSPECTIVE

RALF BLOSSEY

CRC Press
Taylor & Francis Group
Boca Raton London New York

CRC Press is an imprint of the
Taylor & Francis Group, an **informa** business

A CHAPMAN & HALL BOOK

CRC Press
Taylor & Francis Group
6000 Broken Sound Parkway NW, Suite 300
Boca Raton, FL 33487-2742

First issued in paperback 2019

ISBN-13: 978-1-58488-556-6 (hbk)
ISBN-13: 978-0-367-39082-2 (pbk)

**Visit the Taylor & Francis Web site at
http://www.taylorandfrancis.com**

**and the CRC Press Web site at
http://www.crcpress.com**

To SyMaNi and Mum

Preface

This is not a biophysics book.

The readers of this book will indeed find a number of topics in it which commonly classify as biophysics; an example is the discussion of the electrostatic properties of biomolecules. But the book's ambition is different from being yet another introduction into biophysics. I therefore like to explain my motivation for the selection of topics I made.

The title of the book establishes a link between computational biology on the one side, and statistical mechanics on the other. Computational biology is the name of a new discipline. It is probably fair to say that it is, in fact, an emerging one, since a new name is not sufficient to establish a new discipline. Quantitative methods have been applied to biological problems for many years; this has led to a vast number of different subdisciplines of established fields of science: there is a mathematical biology, a biomathematics, a biostatistics, a bioinformatics, a biophysics, a theoretical biology, a quantitative biology... this list is certainly not exhaustive.

All these subdisciplines emerged out of the existing fields and developed by an act of transfer: the use of a method or a mathematical approach within a new context, its application to new problems. One may expect that at some point in the future all these different subdisciplines may merge into a common discipline. For the time being and for the lack of a definite name, let us call this discipline computational biology.

This book wants to contribute a particular element to this field; the use of statistical mechanics methods for the modelling of the properties of biological systems. Statistical physics is the scientific discipline which was developed in order to understand the properties of matter composed of many particles. Traditionally, it has been applied to non-living matter: gases, liquids, and solids. Meanwhile, it is increasingly applied to what is nowadays called 'soft matter' which encompasses complex objects like colloids, membranes and biomolecules, hence objects which do not clearly fall into any one of the classic categories.

Statistical mechanics methods are therefore indeed essential in their application to biophysical problems, since they are needed to understand the static and dynamic properties of biomolecules, complex molecular machines and even whole cell behaviour.

But there is a second aspect for which these methods can prove important, and this relates to the information content of the biological systems. Biology is built on recognition processes: DNA strands have to recognize each other, proteins have to identify DNA binding sites etc. In bioinformatics, these recognition problems are commonly modelled as pattern recognition problems: this mapping is the basis of the enormous success of the field of modern genomics.

Moving beyond genomics, however, to the biology of whole systems, it becomes increasingly clear that an understanding of the physical properties of biological systems becomes more and more important for processes involving biological information content: DNA is not merely a string spelled out in a four-letter alphabet, but it is also an elastic string. There is biological information contained in its *structure* and its *dynamics*. Biology employs physical mechanisms to organize its information processes. This becomes particularly evident, as we begin to understand, in the properties of chromatin, the DNA-protein complex making up the chromosomes in the nuclei of eukaryotic cells (a topic which is discussed in the book). Or, in the particle interaction networks upon which the cellular machinery relies.

This book is placed at just this interface: between biological recognition on the one hand, and the statistical physics methods that can be employed to understand the underlying mechanisms.

The first part of the book gives a concise introduction into the main concepts of statistical mechanics, equilibrium and nonequilibrium. The exposition is introductory in the choice of topics addressed, but still mathematically challenging. Whenever possible, I have to tried to illustrate the methods with simple examples.

The second part of the book is devoted to biomolecules, to DNA, RNA, proteins, and chromatin - i.e., the progression of topics follows more or less what is commonly known as the central dogma of molecular biology. In this part, mostly equilibrium statistical mechanics is needed. The concern here is to understand and model the processes of base-pair recognition and (supra-) molecular structure formation.

The third part of the book is devoted to biological networks. Here, both equilibrium and nonequilibrium concepts introduced in the first part are used. The presentation covers several of the nonequilibrium statistical physics approaches described in Chapter 2 of Part I, and illustrates them on biologically motivated and relevant model systems.

Throughout the book, *Exercises* and *Tasks* are scattered, most of them in the first part. They are intended to motivate the readers to participate actively in the topics of the book. The distinction between *Exercises* and *Tasks* is the following: *Exercises* should be done in order to verify that the concept that was introduced has been understood. *Tasks* are more ambitious and usually require either a more involved calculation or an additional idea to obtain the answer.

A final technical note: the book is complemented by a detailed index. Keywords listed are marked in italics throughout the text.

In the course of shaping the idea of this book and writing it, I profited from discussions with many colleagues. I am especially thankful to Arndt Benecke, Dennis Bray, Martin Brinkmann, Luca Cardelli, Enrico Carlon, Avi Halperin, Andreas Hildebrandt, Martin Howard, Oliver Kohlbacher, Hans Meinhardt, Thomas Lengauer, Hans-Peter Lenhof, Annick Lesne, Ralf Metzler, Johan Paulsson, Andrew Phillips, Wilson Poon, Helmut Schiessel, Bernard Vandenbunder, Jean-Marc Victor, Pieter Rein ten Wolde, Edouard Yeramian.

Finally, I particularly like to thank Andreas Hildebrandt and Martin Howard for their detailed and helpful remarks on an early version of the manuscript.

I gratefully acknowledge the hospitality of the Institut d'Électronique, de Microélectronique et de Nanotechnologie (IEMN) in Villeneuve d'Ascq and the Institut des Hautes Études in Bures-sur-Yvette, where parts of this book were written.

Contents

III   Networks   **183**

Part I

Statistical Mechanics

Chapter 1

Equilibrium Statistical Mechanics

*Après tout, vous êtes ici, c'est l'essentiel! Nous allons faire du bon travail ensemble [...]! D'ici peu, le monde ébloui decouvrira la puissance du grand **Z**!*

Franquin, Z comme Zorglub (1961)

1.1   Z: The partition function

This section introduces the basic physical concepts and mathematical quantities needed to describe systems composed of *many particles*, when only a statistical description remains possible.

Already for an apparently trivial system such as a gas of identical atoms, composed of $N \sim 10^{23}$ particles/mole (Avogadro's number), the description of each particle's trajectory is illusory and not even desirable, and hence only a statistical approach feasible. In biology, as we will see, things are more complicated: not only is the number of relevant interacting components large, but in addition the components are very often 'individuals'.

Suppose we can characterize our system of interest by a given number of 'state variables', whose number can be finite or infinite. An example of such a state variable can be the spatial position of a particle, hence a continuous variable, but in principle it can be any other of its distinguishing characteristics. We call the set of these state variables $\mathbf{x}$, irrespective of their physical nature, and treat $\mathbf{x}$ here as a set of discrete variables;[1] we call such a collection of variables characterizing the state of a system a *microstate*. The possible microstates a system can assume, typically under certain constraints, can be subsumed under the name of a *statistical ensemble*. One has to distinguish the

---

[1] In the case of continuous variables, sums have to replaced by integrals in an obvious way; we will encounter this later and will pass liberally from one notation to the other.

microstate the system can assume in each of its realisations from the ultimate *macrostate* that is to be characterized by a macroscopic physical observable after a suitable averaging procedure over the microstates.[2]

Example. We give a simple illustration of the concept by considering the conformations of a linear molecule such as DNA. The macrostate is the chain conformation we will most probably observe when we image the molecule under certain experimental conditions, while a microstate is any mechanically possible conformation of the molecule. The collection of all these possible states is what we call the statistical ensemble.

Coming back to the conceptual question: what quantity governs the probability $P(\mathbf{x})$ of each state to occur?

Probability and information. We call $P(\mathbf{x}) \geq 0$ a *probability*; consequently, $\sum_{\{\mathbf{x}\}} P(\mathbf{x}) = 1$, where the summation is over all possible states. To each probability we can associate an information measure[3]

$$s(\mathbf{x}) = -\ln P. \tag{1.1}$$

For $P = 1$, $s = 0$, and for $P = 0$, $s = \infty$. Therefore $s(\mathbf{x})$ is a measure for the *lack of information*. It can be used to define *entropy* as the average of $s(\mathbf{x})$ in distribution,

$$S(P) = \sum_{\{\mathbf{x}\}} s(\mathbf{x})P(\mathbf{x}) = -\sum_{\{\mathbf{x}\}} P(\mathbf{x}) \ln P(\mathbf{x}). \tag{1.2}$$

In our terminology entropy can also be called an *average ignorance*. The information-theoretic unit of entropy is the bit.

Example. Again we take as a simple example DNA, which consists of the four bases A, C, G, T (the details of the build-up of DNA are explained in Part II of the book). If the probability of all bases is identical, $P(base) = 1/4$, and we have $S(P) = 4 \times (1/4) \times \ln_2(4) = 2$. The average ignorance is thus 2 bits per nucleotide. In order to interpret this result consider the following: before a base is read by some device (a polymerase transcribing DNA into RNA, see Part III), the information can be represented in binary code as two bits: 11, 10, 01, 00. After reading, the uncertainty has become 0.

---

[2]The averaging procedure means that we can determine the macroscopic state just by an average over the ensemble of microstates; the resulting *ensemble average* gives a correct description of the macrostate of the system if the system had sufficient time in its dynamic evolution to sample all its microstates; this condition is called *ergodicity*.

[3]The choice for the logarithm will become clear in the section on thermodynamics. Note that we do not distinguish in our notation between log and ln, the meaning should be clear from the context.

Maximum entropy. What determines the form of P? In order to deduce it, we employ the prescription proposed by E. T. JAYNES, 1957. It states that the *prior probability distribution* maximizes entropy while respecting macroscopic constraints; it should thus yield the *maximum average ignorance* under those constraints.

Consider first the case of equiprobabilities, as we have assumed for our simple DNA example: there is no other constraint. In order to apply Jaynes' concept, we have to maximize $S(P)$ under the only natural constraint of normalization, $\sum_{\{\mathbf{x}\}} P(\mathbf{x}) = 1$. This leads to a variational problem[4]

$$\delta[S + \lambda[\sum_{\{\mathbf{x}\}} P(\mathbf{x}) - 1]] = 0 \tag{1.3}$$

in which δ denotes the *variation* of the bracketed term with respect to P; e.g., one has $\delta S(P) = -\sum_{\{\mathbf{x}\}}[\delta P \cdot \ln P + P \cdot \delta(\ln P)]$, and $\delta \ln P = [d(\ln P)/dP]\delta P$. In eq.(1.3), λ is a *Lagrange multiplier*, a constant to be determined in the calculation.

This variational problem eq.(1.3) is solved by (*Exercise!*)

$$P(\mathbf{x}) = e^{\lambda - 1} = const. \equiv \Omega^{-1} \tag{1.4}$$

where Ω is the number of realizations of $\mathbf{x}$. All states are indeed found to be equally probable, with a probability inverse to their number. From the definition of entropy, eq.(1.2), we have the relation

$$S = \ln \Omega. \tag{1.5}$$

The canonical ensemble. We now repeat the above argument and calculation for the so-called *canonical ensemble*. Here, one additional constraint appears, since we want to characterize the states of the system now by an *energy* $E(\mathbf{x})$, which we allow to fluctuate with average $\langle E \rangle = E_0$.

In this case we must maximize $S(P)$ under two constraints, the normalization condition, and in addition the condition on the average energy

$$\langle E \rangle = \sum_{\{\mathbf{x}\}} E(\mathbf{x})P(\mathbf{x}). \tag{1.6}$$

---

[4]For those confused by the δ-notation common to variational calculations, the same result is obtained by replacing $P \to P_0 + \epsilon \times P$ with a small parameter ϵ, and requiring the terms in linear order in ϵ to vanish.

The resulting variational problem involves two Lagrange parameters, λ and β:

$$\delta[S + \lambda[1 - \sum_{\{\mathbf{x}\}} P(\mathbf{x})] + \beta[\langle E \rangle - \sum_{\{\mathbf{x}\}} E(\mathbf{x})P(\mathbf{x})]] =$$

$$= \sum_{\{\mathbf{x}\}} \delta P(\mathbf{x}) \cdot (-\ln P - 1 - \lambda - \beta E(\mathbf{x})) = 0 \,, \tag{1.7}$$

with the *canonical* or *Gibbs distribution* as a result,

$$P_\beta(\mathbf{x}) = Z_\beta^{-1} e^{-\beta E(\mathbf{x})} \,, \tag{1.8}$$

where Z_β is the *partition function*

$$Z_\beta = \sum_{\{\mathbf{x}\}} e^{-\beta E(\mathbf{x})} \,. \tag{1.9}$$

For dimensional reasons, the Lagrange parameter β must be an inverse energy, which we take as the *thermal energy*

$$\beta^{-1} = k_B T \tag{1.10}$$

with Boltzmann's constant k_B. We define the *free energy* as

$$F \equiv -k_B T \ln Z \,. \tag{1.11}$$

The meaning of this definition will become clear in the following section. Eq.(1.9) is the key quantity in all of the first chapter of this part of the book; we will be mostly concerned with methods how to compute it.

Task. Determine $P_{\beta,\mu}$ if not only the average energy $\langle E \rangle$ is taken as a constraint condition, but also the number of particles N is allowed to fluctuate with average $\langle N \rangle$. The Lagrange parameter μ is called *chemical potential*.[5] The ensemble governed by the resulting distribution is called the *grand canonical ensemble*.

Equivalence of ensembles. The different ensembles (canonical, grand canonical etc.) are macroscopically equivalent for equilibrium states in the *thermodynamic limit* of N, $V \to \infty$ with $n = N/V = const$. Deviations from the thermodynamic limit scale as $\sim 1/\sqrt{N}$, and are hence negligible since the value of N we usually talk about in statistical mechanics is on the order of Avogadro's constant.[6] This equivalence allows to choose the ensemble based

---

[5]The chemical potential often confuses: it is a measure for the availability of particles, and plays thus an analogous role as temperature does in providing an energy source.

[6]There are exceptions for the equivalence of ensembles, a point we do not pursue here. And, obviously, for systems in which N is *very much smaller* than Avogadro's constant, the deviations from the thermodynamic limit can be important.

on its convenience in a particular application - we will encounter such cases in the second part of the book.

1.2 Relation to thermodynamics

The knowledge of the statistical distributions for the different ensembles is sufficient to characterize the macroscopic properties of a physical system in thermal equilibrium. We now make the link explicit between the statistical description and the expressions of macroscopic thermodynamics, for which a number of basic principles can be formulated.

The Laws of Thermodynamics. The physical insight of macroscopic thermodynamics is usually summarized in the *laws of thermodynamics*. They are

- Law **0**: If two systems A and B are in *thermodynamic equilibrium*, and B is in equilibrium with C, then A and C are also in equilibrium with each other. The equilibria can be characterized by their mechanical, thermal or chemical properties.

- Law **1**: Energy E is conserved:

$$dE = dQ + dW \qquad (1.12)$$

  where Q is the *heat* flowing into the system, and W the *work* done by the system. If a process is *adiabatic*, dQ is zero.

- Law **2**: In a thermally isolated macroscopic system, entropy never decreases.

- Law **3**: As the temperature of a system tends to absolute zero, entropy reaches a constant.

We see that energy and entropy are the key notions of equilibrium thermodynamics. Let us understand them more deeply. To achieve this we first generalize the descriptors we have introduced before and introduce the important notion of an *extensive* variable.

An extensive variable is a variable which scales linearly with system size. Volume is a (trivial) example for such a system property: if one starts with two systems of volume V_1 and V_2, the volume of the systems is additive: $V = V_1 + V_2$, and hence scales linearly. The particle number N is another example. Temperature T, however, is not extensive: putting two systems of equal temperature T together does not result in a system with temperature $2T$.

We call the state of a system an *equilibrium state* if it can be characterized by its (internal) energy E, and a set of extensive parameters $X_1, ... X_m$. For such a state there exists a function of the extensive parameters, the entropy S (which we obtained before from the Jaynes principle). We call the relation

$$S = S(E, X_0, ..., X_m) \tag{1.13}$$

a fundamental equation for the system. S itself is also extensive; furthermore, it is an increasing function[7] of E.

Consider now the first statement, the notion of the extensivity of S. We can express it mathematically by writing S as a *first-order homogeneous* function of its extensive parameters

$$S(\lambda E, \lambda X_0, ... \lambda X_m) = \lambda S(E, X_0, ..., X_m), \tag{1.14}$$

where λ is a scalar.

The second statement about S, the monotonicity property, corresponds to the condition[8]

$$\left(\frac{\partial S}{\partial E} \right)_{X_i} \geq 0. \tag{1.15}$$

The monotonicity property of S with respect to E allows its inversion, leading to

$$E = E(S, X_0, ..., X_m) \tag{1.16}$$

and E, like S, is a first-order homogeneous function

$$E(\lambda S, \lambda X_0, ... \lambda X_m) = \lambda E(S, X_0, ..., X_m). \tag{1.17}$$

Some properties of scalar fields. Before introducing the thermodynamic potentials, we need some mathematical concepts for scalar fields. They are stated here just as facts.

---

[7] E is also extensive, but this is actually a subtle point if systems become strongly interacting. We are not concerned with sophisticated problems of this sort here.
[8] The equality is restricted to a set of measure zero.

- If $\phi = \phi(x_0, ..., x_m)$ is a scalar field of $m+1$ variables, its *total differential* is given by

$$d\phi = \sum_{i=0}^{m} \frac{\partial \phi}{\partial x_i} dx_i. \tag{1.18}$$

- If the $x_i = x_i(u, v)$ with u, v scalar, we can rewrite this expression as

$$d\phi = \sum_{i=0}^{m} \frac{\partial \phi}{\partial x_i} \frac{\partial x_i}{\partial u} du + \sum_{i=0}^{m} \frac{\partial \phi}{\partial x_i} \frac{\partial x_i}{\partial v} dv. \tag{1.19}$$

- *Contour surfaces*, for $\phi = constant$, define an implicit functional relationship between the x_i since on a contour surface

$$\sum_{i=0}^{m} \frac{\partial \phi}{\partial x_i} dx_i = 0. \tag{1.20}$$

- If all x_i except x_0 and x_1 (e.g.) are held fixed, then

$$\left(\frac{\partial x_1}{\partial x_0}\right)_{\phi,...} = -\left(\frac{\partial \phi}{\partial x_0}\right)_{x_1,...} \cdot \left(\frac{\partial \phi}{\partial x_1}\right)_{x_0,...}^{-1}, \tag{1.21}$$

and

$$\left(\frac{\partial x_0}{\partial x_1}\right)_{\phi,...} = \left(\frac{\partial x_1}{\partial x_0}\right)_{\phi,...}^{-1} \tag{1.22}$$

for $\phi = const$.

- For three variables, one has the cyclic rule

$$\left(\frac{\partial x_0}{\partial x_1}\right)_{x_2} \cdot \left(\frac{\partial x_1}{\partial x_2}\right)_{x_0} \cdot \left(\frac{\partial x_2}{\partial x_0}\right)_{x_1} = -1, \tag{1.23}$$

which generalizes to more variables in an obvious way (*Exercise*).

We can now proceed to look at the total differentials of E and S. We assume that E is a function of entropy S, volume V, and particle number N; the extension to further extensive variables is straightforward. We have

$$dE = \left(\frac{\partial E}{\partial S}\right)_{V,N} dS + \left(\frac{\partial E}{\partial V}\right)_{S,N} dV + \left(\frac{\partial E}{\partial N}\right)_{S,V} dN \tag{1.24}$$

where the subscripts indicate the variables to be held fixed. We define the following intensive parameters:

$$T = \left(\frac{\partial E}{\partial S}\right)_{V,N} \tag{1.25}$$

is *temperature*;

$$P = -\left(\frac{\partial E}{\partial V}\right)_{S,N} \tag{1.26}$$

is *pressure*;

$$\mu = \left(\frac{\partial E}{\partial N}\right)_{S,V} \tag{1.27}$$

is *chemical potential*; hence

$$dE = TdS - PdV + \mu dN \,. \tag{1.28}$$

The intensive parameters are all functions of S, V, N, and the functional relationships in eqs.(1.25) - (1.27) are called *equations of state*.

Exercise. Verify explicitly that temperature T, pressure P and chemical potential are *intensive variables*, i.e., that they are homogeneous functions of zeroth order.

Exercise. Deduce from the fact that E is a first-order homogeneous function of its variables one can obtain the relation (the *Euler equation*[9])

$$E = TS - PV + \mu N \,. \tag{1.29}$$

Hint: differentiate eq.(1.17) with respect to λ and put $\lambda = 1$.

Given that we have introduced E as a function of entropy S, volume V and particle number N, one may wonder whether it is not possible - and even desirable - to define thermodynamic functions of other variables than the ones chosen. In particular, one may also want to study dependences on intensive rather than extensive variables. This is indeed possible, and the resulting functions are not independent from each other. In fact, one can pass from one to the other via a mathematical transformation we introduce first in a formal manner.

Legendre transform. Let $Y(X_0, ..., X_m)$ be a scalar field of the extensive variables X_j, and the $P_j = (\partial Y/\partial X_j)_{X_{i\neq j}}$ are the corresponding intensive variables. Then

$$\Lambda = Y[P_0, ..., P_i]_{i\leq m} \equiv Y - \sum_{j=0}^{i} X_j P_j \tag{1.30}$$

---

[9]The expressions for E show that TS has the dimension of an energy. Hence the information entropy we determined in the first section has to be multiplied by Boltzmann's constant, k_B.

is the *Legendre transform* of Y with respect to $X_{j \leq i}$. The total differential of Λ reads

$$d\Lambda = -\sum_{j=0}^{i} X_j dP_j + \sum_{j=i+1}^{m} P_j dX_j \tag{1.31}$$

so that $\Lambda = \Lambda(P_0, ..., P_i, X_{i+1}, ..., X_m)$ is a function of the $i+1$ intensive, and $m - i$ extensive variables.

After this formal definition we want to apply this concept. This is best done by simple cases, ignoring the physical context for the moment.

Legendre transform in one dimension. We first discuss the Legendre transform in one dimension for a function $f(x)$ and its derivative

$$y = f'(x) = \frac{df}{dx} \equiv g(x). \tag{1.32}$$

We can understand $y = g(x)$ as a variable transformation from x to y. How can we express f in terms of y?

For the function $f(x)$ we can write the equation within each point x along the curve

$$f(x) = x f'(x) + b(x). \tag{1.33}$$

Since $x = g^{-1}(y)$ where g^{-1} is the inverse of g, we have

$$b(g^{-1}(y)) = f(g^{-1}(y)) - y g^{-1}(y) \equiv \Lambda(y). \tag{1.34}$$

This is often written in short form as

$$\Lambda(y) = f(x) - xy, \tag{1.35}$$

where x is a function of y.

We move on to some exercises.

Exercise. Compute the Legendre transform of $f(x) = e^x$.

Exercise. Give the Legendre transform of the harmonic form with the vector **x**,

$$u(\mathbf{x}) = \frac{1}{2}(\mathbf{x}^T \cdot \mathbf{A} \cdot \mathbf{x}), \tag{1.36}$$

where $\mathbf{A}$ is an invertible $(n \times n)$-matrix.

We now return to the physical context. For the energy E, the four most common thermodynamic potentials that result from the application of the Legendre transform are the

- *Helmholtz free energy* (T given, canonical ensemble)

$$F(T, V, N) = E[T] = E - TS = -PV + \mu N \qquad (1.37)$$

$$dF = -SdT - PdV + \mu dN \qquad (1.38)$$

- *Enthalpy* (P given)

$$H(S, P, N) = E[P] = E + PV = TS + \mu N \qquad (1.39)$$

$$dH = TdS + VdP + \mu dN \qquad (1.40)$$

- *Gibbs free energy* (T, P given)

$$G(T, P, N) = E[T, P] = E + PV - TS = \mu N \qquad (1.41)$$

$$dG = -SdT + VdP + \mu dN \qquad (1.42)$$

- *Grand canonical potential* (T, μ given)

$$\Phi(T, V, \mu) = E[T, \mu] = E - TS - \mu N = -PV \qquad (1.43)$$

$$d\Phi = -SdT - PdV - Nd\mu \qquad (1.44)$$

The thermodynamic potentials all have to be minimized at equilibrium for fixed values of their variables.

This ends our brief look into thermodynamics. The message that we want to retain is that

- within equilibrium physics there is a well-established body of thermodynamic functions with which the macroscopic properties of a system can be described;

- these functions have a rigorous link with each other via the Legendre transform, and with statistical mechanics, since we know how to compute the thermodynamic potentials within this theory.

We now return to statistical mechanics, and begin to discuss methods which allow to compute the partition function and the quantities derivable from it.

1.3 Computing Z

This section introduces methods to compute Z and the thermodynamic quantities that can be derived from it. We begin with some technicalities that will be useful later. An obvious first step in the computation of Z is to know how to compute integrals involving the *Gaussian distribution*. The Gaussian distribution is a generic characteristic of equilibrium states: in thermal equilibrium, a system will be in a state minimizing the corresponding thermodynamic potential, and the fluctuations around this stable state will be Gaussian.

Gaussian distribution. The Gaussian probability distribution in the case of one variable $-\infty < x < \infty$ is given by

$$P(x) = C e^{-\frac{1}{2}Ax^2 - Bx} \tag{1.45}$$

where the normalization constant is

$$C = \left(\frac{A}{2\pi}\right)^{1/2} e^{-\frac{B^2}{2A}} . \tag{1.46}$$

The parameter $A > 0$ controls the width of $P(x)$ and, together with B, the peak position. Introducing the average μ and variance σ^2 we find

$$\mu_1 = -\frac{B}{A}, \quad \sigma^2 = \frac{1}{A} \tag{1.47}$$

and can thus write the normalized form of the Gaussian or, *standard normal distribution*

$$P(x) = \frac{1}{\sqrt{2\pi\sigma^2}} e^{-\frac{(x-\mu_1)^2}{2\sigma^2}} . \tag{1.48}$$

The *multivariate* version of the distribution, for $i = 1, ..., m$ random variables $x = \{x_i\}$ is

$$P(x) = C \exp\left[-\frac{1}{2}\sum_{i,j=1}^{m} A_{ij}x_i x_j - \sum_{i=1}^{m} B_i x_i\right] \tag{1.49}$$

where $\mathbf{A} = A_{ij}$ is a positive definite symmetric matrix. The normalization constant reads in this case as

$$C = (2\pi)^{-m/2} (Det\,\mathbf{A})^{-1/2} \exp\left[-\frac{1}{2}\mathbf{B}^T \cdot \mathbf{A}^{-1} \cdot \mathbf{B}\right] \tag{1.50}$$

with the matrix inverse A^{-1}.

Based on these results we can now easily write down the *mean* of each of the x_i

$$\langle x_i \rangle = -\sum_j (A^{-1})_{ij} B_j \tag{1.51}$$

and the *covariance*

$$\langle (x_i - \langle x_i \rangle)(x_j - \langle x_j \rangle) \rangle = \langle x_i x_j \rangle - \langle x_i \rangle \langle x_j \rangle = (A^{-1})_{ij} \tag{1.52}$$

The inverse of A is the *covariance matrix*. Its diagonal elements are the variances, the off-diagonals are the co-variances.

Characteristic function, moment generating function and cumulants. The *characteristic function* of a stochastic variable X is defined by

$$G(k) \equiv \langle e^{ik} \rangle = \int dx P(x) e^{ikx} , \tag{1.53}$$

where the symbol $\langle ... \rangle$ was used to abbreviate the average in distribution. Obviously, $G(k)$ is a Fourier transform of P. It exists for real k and obeys

$$G(0) = 1 , \quad |G(k)| \leq 1 . \tag{1.54}$$

The coefficients μ_m of its Taylor expansion are the *moments*

$$G(k) = \sum_{m=0}^{\infty} \frac{(ik)^m}{m!} \mu_m \tag{1.55}$$

where

$$\mu_n = \langle x^n \rangle = \int dx P(x) x^n . \tag{1.56}$$

The coefficients κ_m of the series of its logarithm,

$$\ln G(k) = \sum_{k=1}^{\infty} \frac{(ik)^m}{m!} \kappa_m \tag{1.57}$$

are the *cumulants*. They are combinations of the moments, the lowest of which are

$$\kappa_1 = \mu_1 , \quad \kappa_2 = \mu_2 - \mu_1^2 = \sigma^2. \tag{1.58}$$

For the Gaussian distribution, all higher cumulants vanish. Higher moments can thus serve to characterize more complex distributions - this is useful since not in all cases full distributions are obtainable.

Exercise. Compute the cumulants for the *Poisson distribution*

$$p_n = \frac{a_n}{n!} e^{-n} \tag{1.59}$$

over the integers and zero, $n = 0, 1, 2, ...$

We have reached a point where we want to see the machinery of statistical mechanics in action. We will now apply it to two standard model systems of statistical mechanics. The first is the *Ising model*, probably the most famous model ever formulated in statistical mechanics. Originally invented to describe an uniaxial ferromagnet and its phase transition from the paramagnetic (non-magnetic) to the ferromagnetic phase, it has found innumerable applications all over the fields of physics, chemistry and biology.

The second model we will introduce is that of the mechanics of a polymer. Again think of DNA, but now on a different length scale, beyond that of the base pairs. The molecule is considered as an elastic string which resists bending in space, but which can also be pulled along its axis.

We begin by a discussion of the one-dimensional Ising model and its solution.

The 1d-Ising model. The Ising model on a one-dimensional lattice is defined by the energy[10]

$$H = -\frac{K}{2} \sum_n s_n s_{n+1} - h \sum_n s_n \tag{1.60}$$

where the $s_n = \pm 1$ are called the 'spins' - local magnetic moments - of a ferromagnet.[11] Within the model, the parameter K describes a coupling between neighbouring local spins; for $K > 0$ they will prefer to be in the same state since the energy favors that state. The parameter h in the second term defines an external coupling affecting each of the spins; for the ferromagnetic system for which the Ising model was originally conceived, this contribution represents an applied magnetic field which, depending on its sign, can favour

---

[10] In statistical mechanics jargon, the *Hamiltonian.*

[11] A ferromagnet is usually a metal which, below a certain temperature, has a permanent magnetic moment; the origin of this permanent moment lies in the microscopic magnetic moments produced by the electron shells inside the metal. In the model description we employ, we summarize these atomistic effects by an elementary spin. This is sufficient if we are only interested in the ordering phenomenon between a paramagnetic (non-magnetic) and a ferromagnetic phase. A true microscopic theory of ferromagnetism has to be a quantum theory.

either the 'up' or the 'down' state with all spins following the field direction.

Since we will use the Ising model in a biological context, it is clear that the original interpretation of the parameters is largely immaterial for us: for us, s_n is just a two-valued variable, and can be re-interpreted at will. From a biological point of view we can, e.g., consider the Ising model in one spatial dimension as a chain of objects which can be in either of two states. A simple biological example to which this model then applies can be a DNA molecule. Within the Ising model we can define microstates that distinguish each other by the different numbers of spins up - interpreted as 'base pairs bound' and spins down, interpreted as 'base pairs unbound'. Hence we can, e.g., address the question of the binding of two DNA strands, and the Ising model becomes a first - but ultimately too crude - model for the physics of a DNA double strand.

This re-interpretability of the model is a big advantage of statistical mechanics: indeed, very often physically quite different situations turn out to fall into the class of just one type of model. This explains the success of the Ising model in so many applications - having a system with just two alternatives is the simplest one can have.

Plugging the Hamiltonian of the Ising model into the partition function in the canonical ensemble, we obtain

$$Z = \sum_{\{s_n\}} e^{-H} = \sum_{\{s_n\}} e^{J \sum_n s_n s_{n+1} + \beta h \sum_n s_n} \qquad (1.61)$$

with $J \equiv \beta K/2$. In the following we leave out the field h; its inclusion in the calculation is left as a *Task*.

Now we have to compute Z, and we will learn a first method to do it.

The transfer-matrix solution of the 1d-Ising model. The partition function can be rewritten as

$$Z = \sum_{\{s_n\}} e^{J s_0 s_1} \cdot e^{J s_1 s_2} \cdot \ldots \cdot e^{J s_{N-1} s_0} \qquad (1.62)$$

where a *periodic boundary condition* is assumed, i.e., the linear chain is joined at its ends.

Since each spin has two orientations, there are four configurations for two neighbouring spins, two of which are degenerate in energy. This suggests to introduce a matrix representation with a *transfer matrix*

$$T_J = \begin{pmatrix} e^J & e^{-J} \\ e^{-J} & e^J \end{pmatrix} \tag{1.63}$$

which is a symmetric matrix, to be diagonalized by an orthogonal matrix O. Since this matrix applies to every neighbouring pair of spins we can express the partition function as (*Exercise*)

$$Z = Tr[T_J^N] = Tr[(OT_JO^{-1})^N] \tag{1.64}$$

where the symbol Tr stands for 'Trace', the sum of the diagonal entries of the matrix. The orthogonal matrix O to diagonalize T_J is given by

$$O = \begin{pmatrix} 0 & 1 \\ -1 & 0 \end{pmatrix} \tag{1.65}$$

and we find for the diagonalized matrix the result

$$OT_JO^{-1} = 2 \begin{pmatrix} \cosh J & 0 \\ 0 & \sinh J \end{pmatrix} \tag{1.66}$$

so that

$$Z = 2^N [\cosh^N J + \sinh^N J] . \tag{1.67}$$

Since in statistical mechanics we are interested in the thermodynamic limit which in this model amounts to consider $N \rightarrow \infty$, one finally finds the expression of the partition function

$$Z = 2^N \cosh^N J . \tag{1.68}$$

Show this as an *Exercise*.

Eq.(1.68) is the first partition function we have calculated explicitly, and we can now calculate physical quantities from it. Let us begin with the internal energy E; it is found to be

$$E = Z^{-1} \sum_{\{s_n\}} He^{-\beta H} = Z^{-1}\frac{\partial Z}{\partial \beta} = -NK \tanh(\beta K/2) . \tag{1.69}$$

A second, more complex example is the two-point correlation function $\langle s_i s_j \rangle$ for $|i - j| = r$. It is given by

$$\langle s_i s_j \rangle = Z^{-1} \sum_{\{s_n\}} s_i s_j e^{-\beta H} . \tag{1.70}$$

In order to apply the transfer matrix method we first have to find a suitable way to represent it in terms of T_J. One can convince oneself that the expression to write down is

$$\sigma_3 T_J^r \sigma_3 T_J^{N-r} \tag{1.71}$$

where σ_3 is the matrix

$$\sigma_3 = \begin{pmatrix} 1 & 0 \\ 0 & -1 \end{pmatrix} . \tag{1.72}$$

This expression states that between two well-defined spin states along the chain the transfer matrix has to propagate r times between the spins, and then again $N - r$ times along the other side of the closed chain - remember we maintain the periodic boundary condition.

With this we can now write

$$\langle s_i s_j \rangle = Z^{-1} Tr[\sigma_3 T_J^r \sigma_3 T_J^{N-r}]$$

$$= Z^{-1} Tr[(O\sigma_3 O^{-1}) \cdot (OT_J O^{-1})^r \cdot (O\sigma_3 O^{-1}) \cdot (OT_J O^{-1})^{N-r}]$$

$$= 2^N Z^{-1} (\cosh^r J \sinh^{N-r} J + \sinh^r J \cosh^{N-r} J)$$

$$= \tanh^{N-r} J + \tanh^r J \tag{1.73}$$

where

$$O\sigma_3 O^{-1} = \begin{pmatrix} 0 & 1 \\ 1 & 0 \end{pmatrix} . \tag{1.74}$$

With $N \gg r$ and in the thermodynamic limit we obtain the expression

$$\langle s_i s_j \rangle = \tanh^r J = e^{-|i-j|/\xi} \tag{1.75}$$

where we defined

$$\xi \equiv |\ln(\tanh J)|^{-1} \tag{1.76}$$

as the *correlation length*. For the 1d-Ising model ξ diverges in the limit $T \to 0$, when all spins align either up or down (i.e., the system adopts a homogeneous, oriented state). It will turn out that there is no *phase transition* in the one-dimensional Ising model at a finite temperature ($T > 0$) between a state of finite magnetization - the oriented state in which all spins point collectively either up or down - and a state without a net magnetization, in which the spins point randomly either up or down with a vanishing average.

For this simple example, we are essentially done and the reader can try to extend the model to include the field h, or calculate other thermodynamic quantities to get a feeling for the systems' properties.

But we do not yet want to stop here and take at little deeper look into this system. We have computed thermodynamic quantities in the thermodynamic limit $N \to \infty$. Let us suppose we don't do that - in fact, in some applications of statistical mechanics in biology, this limit may be difficult to reach, and effects of finite system size may be important.

Let us therefore study how the properties of the one-dimensional Ising model are affected upon a change of N. We start from the expression

$$Z[N, J] = Tr[T_J^N] = Tr[(OT_JO^{-1})^N] \tag{1.77}$$

where $OT_JO^{-1} = e^J \cdot \mathbf{1} + e^{-J}\sigma_3$; the N-dependence is now explicit. Suppose we, in a first step, double the lattice size from N to $2N$. With

$$(OT_JO^{-1})^2 = 2(\cosh 2J \cdot \mathbf{1} + \sigma_3) = COT_{\bar{J}}O^{-1} \tag{1.78}$$

where

$$C = 2\sqrt{\cosh 2J}, \quad \bar{J} = \frac{1}{2}\ln(\cosh 2J) \tag{1.79}$$

we arrive at the relation

$$Z[2N, J] = C^N Z[N, \bar{J}]. \tag{1.80}$$

The equations (1.79), (1.80) can be read as *generators of a flow* in the coupling constant J. If we were to iterate the mapping $J \to \bar{J}$ starting from a value $J = 1$, the iterated map will converge to zero (*Exercise*). What can we learn from this observation?

By these very simple considerations we have in fact found the two *fixed points* of a *renormalization group* transformation of the 1d-Ising model - without noticing that such a thing exists. The first fixed point is $J = 0$, which corresponds to $T = \infty$. Physically, the system is then in the fully disordered phase, in which the spins point randomly either up or down. The second fixed point is $J = \infty$ corresponding to $T = 0$, and this is the fully ordered spin state which only exists at zero temperature. Indeed, there is no phase with a finite magnetization $M \equiv \langle s_i \rangle \neq 0$ for $0 < T \leq \infty$, as we mentioned before.

Although we have of course not made a systematic approach to renormalization, there is already something to learn. Suppose we consider the system at a given temperature, and at a given size N. The spins inside the system will have a correlation length ξ over which their orientation (up or down) is correlated. If we now double the system size - what does the correlation do: will it grow stronger or not? It is characteristic of a system *right at* a phase transition that the correlation length is infinite. The system behaves in a collective (or cooperative) fashion. If the temperature I choose is just that critical temperature, all the steps of increasing system size will not affect the correlation length, and it will stay infinite. For the 1d-Ising model, this is just the case at $T = 0$. By contrast, if I deviate ever so slightly from that critical temperature, the step of increasing system size will tend to reduce the correlation length, and if this procedure is repeated over and over again[12] the system will move away to a system of uncorrelated spins, hence a disordered phase. This whole approach, in its systematic version, allows to compute the properties of all equilibrium phase transitions in a unified way.

After this digression on the idea behind the renormalization group, we turn to another approach to compute the partition function of the 1d-Ising model: the computation by recursion. This will turn out to be a very important technique in the second part of the book (Part II: Biomolecules), since recursion methods are at the heart of dynamic programming.

Solution of the 1d-Ising model by recursion. The analytic solution of the 1d-Ising model is so easy to obtain since the coupling between two spins is identical. This is not necessarily the case, and we will indeed see in Part II that for the most relevant biological applications this simplifying assumption can not be made.

So let's now make things a little more complicated. For the 1d-Ising model with a *neighbour-dependent coupling* $J(n)$, we write[13]

$$\beta H = - \sum_{n=1}^{N-1} J(n)s(n)s(n+1) - \sum_{n=1}^{N} H(n)s(n) \qquad (1.81)$$

where we now have also included a neighbour-dependent field $H(n) = \beta h(n)$. This variant of the Ising-model has been applied to problems in statistical genetics, as suggested by J. MAJEWSKI et al., 2001, in a study of epistasis

---

[12]Hence the notion of a *renormalization group*; technically, it is a semi-group.
[13]With a slight change of notation which should be obvious.

(i.e., gene-gene interaction). In this interpretation, the spins on the lattice are identified with different genes placed along the DNA molecule rather than the individual base pairs.

As before, the total number of states in the model is 2^N, hence exponential in N, and the computation to be performed is seemingly exponential in N since we now cannot use the transfer matrix trick anymore due to the neighbour-dependence of couplings. Using a recursion technique, however, we can achieve a computation in linear 'time' N.

For the construction of the recursion relation we write

$$Z_N = Z_{N+} + Z_{N-} \tag{1.82}$$

where Z_{N+} refers to the partition function of the chain (which is left open in this case - in contrast to our computation before) in which the last spin at site N points up; the interpretation of Z_{N-} is evident. One finds (*Exercise*)

$$Z_{N+} = e^{H(N)}[Z_{(N-1)+}e^{J(N-1)} + Z_{(N-1)-}e^{-J(N-1)}] \tag{1.83}$$

and

$$Z_{N-} = e^{-H(N)}[Z_{(N-1)+}e^{-J(N-1)} + Z_{(N-1)-}e^{J(N-1)}]. \tag{1.84}$$

The partition function Z_N can now be computed recursively from these expressions.

This concludes the discussion of the Ising model in one spatial dimension.

Although we saw some calculational approaches at work, we can be a little bit dissatisfied since there is no phase transition between an ordered and a disordered state in this model at a finite temperature. So we are asked to generalize, and one way to do this is to pass on to higher dimensions.[14] Since calculations in higher dimensions are more complicated (and in fact, they will become far too complicated for the ambitions of this text) we will try to simplify in another way. This leads us to what is called the *mean-field solution* of the Ising model. In this approach, spatial dimension will first play no role.

Mean-field solution. For the discussion of the *mean-field approximation* we place the spins on the edges of a hypercube in d space dimensions, and

---

[14]With this generalization we lose for the moment the biological interpretation of the Ising model for the binding of a DNA double strand. But this doesn't matter: here we are mainly interested in the problem of the calculation of partition functions; the more specific use of such models within a given biological context follows in the subsequent parts of the book.

write the energy as

$$H = -\frac{1}{2}\sum_{\mathbf{x},\mathbf{y}} J(\mathbf{x},\mathbf{y})s(\mathbf{x})s(\mathbf{y}) \tag{1.85}$$

where $\mathbf{x} = a\,x_i\,\mathbf{e}_i$ with the lattice constant a and $\mathbf{e}_i$, $i = 1,...,d$ are the unit vectors on the lattice. $J(\mathbf{x},\mathbf{y})$ is taken as a symmetric matrix.

Again we need a technical concept first.

Hubbard-Stratonovich transformation. Consider the Gaussian integral relation

$$\exp\left(\frac{\beta}{2}Js^2\right) = \sqrt{\frac{\beta}{2\pi J}}\int_{-\infty}^{\infty} d\phi\, \exp\left(-\frac{\beta}{2}J^{-1}\phi^2 + \beta\phi s\right). \tag{1.86}$$

It can be understood as a *linearization* of the argument of the exponential (i.e., the variable s). The price to pay for this operation is the introduction of an auxiliary integration variable.

We generalize this step to the weight

$$\exp\left(\frac{\beta}{2}\sum_{\mathbf{x},\mathbf{y}} J(\mathbf{x},\mathbf{y})s(\mathbf{x})s(\mathbf{y})\right) = \tag{1.87}$$

$$= \prod_{\mathbf{x}}\int_{-\infty}^{\infty} d\phi(\mathbf{x})\exp\left[-\frac{\beta}{2}\sum_{\mathbf{x},\mathbf{y}} J^{-1}(\mathbf{x},\mathbf{y})\phi(\mathbf{x})\phi(\mathbf{y}) + \beta\sum_{\mathbf{x}}\phi(\mathbf{x})s(\mathbf{x})\right]$$

where we now have introduced a local *order parameter field* $\phi(\mathbf{x})$ at each lattice site. Note that the expression is meaningful as long as the matrix is semi-definite and positive, which is not the case for the Ising model since $J(\mathbf{x},\mathbf{x}) = 0$; the formula thus has to be considered a formal one.[15]

The partition sum over the spins s can be performed exactly with the Hubbard-Stratonovich transformation, with the result

$$Z = \int D\phi\,\exp\left[-\frac{\beta}{2}\sum_{\mathbf{x},\mathbf{y}} J^{-1}\phi(\mathbf{x})\phi(\mathbf{y}) + \sum_{\mathbf{x}}\ln(\cosh[\beta\phi(\mathbf{x})])\right] \tag{1.88}$$

where the integral measure is $\int D\phi \equiv \int \prod_{\mathbf{x}} d\phi(\mathbf{x})$. The $\ln(\cosh(..))$-term in the argument of the exponential function arises from the summation over the

---

[15]Further, we have ignored the integration amplitude.

spins $s(\mathbf{x})$ which is now easy to do - that was just the idea behind the linearization in the first place.

We now have the expressions in place to perfom the mean-field approximation. It corresponds to taking the *saddle-point* of the integrand, which is determined by the stationary value of the argument of the exponential. This idea is easily explained for a function of one variable; it is also known as the *method of steepest descent*.[16] Suppose the integral we want to compute is

$$I = \lim_{N\to\infty} \int_{-\infty}^{\infty} dx\, e^{-Nf(x)} \tag{1.89}$$

where we have introduced an explicit parameter N; frequently such a parameter can be defined, if often only on formal grounds. Suppose the function f has a global minimum at a value $x = x_0$, well-separated from possibly other minima. Upon Taylor expansion, f fulfills

$$f(x) \approx f(x_0) + \frac{1}{2}f''(x_0)(x-x_0)^2 + \mathcal{O}(x^3)\,, \tag{1.90}$$

since we expand around an extremum for which $f'(x_0) = 0$.

In the limit of $N \to \infty$, the global minimum will dominate the integrand and the integration range is largely determined by the region around x_0. Hence

$$I \approx \lim_{N\to\infty} e^{-Nf(x_0)} \int_{-\infty}^{\infty} dx\, e^{\frac{N}{2}f''(x_0)(x-x_0)^2} \approx \lim_{N\to\infty} e^{-Nf(x_0)} \left(\frac{2\pi}{Nf''(x_0)}\right)^{1/2}. \tag{1.91}$$

Task. Compute the next order correction to the saddle-point value of the integral.

Coming back to our case, the result is given by the equation

$$-\sum_{\mathbf{y}} J^{-1}(\mathbf{x},\mathbf{y})\phi_0(\mathbf{y}) + \tanh[\beta\phi_0(\mathbf{x})] = 0\,. \tag{1.92}$$

with the mean-field $\langle\phi(\mathbf{x})\rangle = \phi_0(\mathbf{x})$.

A note on the mean-field approximation. At this point it is useful to make a comment on the notion of the mean-field approximation since the approach we took seems rather technical. In fact, there are several ways of

---

[16] For oscillating integrands, see the *stationary phase approximation*.

introducing a mean-field. A simple alternative is, e.g., to define and introduce

$$\widehat{H}(\mathbf{x}) \equiv \sum_{\mathbf{y}} J(\mathbf{x}, \mathbf{y}) s(\mathbf{y}) \qquad (1.93)$$

as an effective field acting on the spins $s(\mathbf{y})$. The resulting partition function then becomes one-dimensional and can be solved following the steps we took before. This procedure essentially means that within mean-field theory, one neglects correlations and factorizes $\langle s(\mathbf{x})s(\mathbf{y})\rangle = \langle s(\mathbf{x})\rangle\langle s(\mathbf{y})\rangle$. While this procedure is technically much easier to perform than the computation we did, we will see in the following that our more systematic and general approach will also allow us, in a fairly straightforward sequence of steps, to learn something about the regime of validity of the mean-field approximation.

We now return to eq.(1.92). If we assume that $J(\mathbf{x}, \mathbf{y}) \equiv J$, the system becomes translationally invariant and we can define the uniform *order parameter* M via

$$\phi_0 = 2dJM \qquad (1.94)$$

so that the saddle-point equation is rewritten as

$$M = \tanh(2d\beta JM). \qquad (1.95)$$

The solutions of this equation can be obtained graphically, see Figure 1.1. If $2d\beta JM > 1$, there is a pair of solutions which merge at $M = 0$ at the critical temperature

$$\beta_c^{-1} = 2dJ. \qquad (1.96)$$

Within the mean-field approximation, the Ising model thus displays a phase transition at a finite temperature. This is at variance with our exact result for $d = 1$. So we have to clarify the limit of validity of the approximation we made.

In fact, as we said before, the mean-field approximation assumes a factorization of correlation functions, hence a neglect of fluctuations.[17] These can be accounted for by an expansion around the saddle-point value. We may thus speculate that the approximation is valid whenever this expansion is meaningful. In order to approach this question, we therefore have to pass on to a theory which is capable to capture the properties of the Ising model in the vicinity of the critical temperature.

Ginzburg-Landau theory. In the vicinity of the transition, we can expand the terms in eq.(1.88) in a Taylor series (Why? *Exercise!*)

---

[17]Why do fluctuations destroy the factorization of correlation functions? *Exercise.*

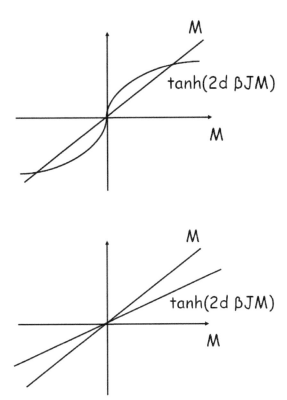

FIGURE I.1.1: Phase transition in the Ising model in the mean-field approximation: graphical solution of the mean-field equation. A change in temperature allows to pass from a unique solution for $T > T_c$ (bottom) to two solutions at $T < T_c$ (top).

$$\sum_{\mathbf{x},\mathbf{y}} J^{-1}\phi(\mathbf{x})\phi(\mathbf{y}) = J^{-1}\sum_{\mathbf{x}} \phi(\mathbf{x})\left(\frac{1}{2d} - \frac{1}{4d^2}a^2\nabla^2 + ...\right)\phi(\mathbf{x}) \qquad (1.97)$$

and

$$\ln[\cosh(\beta\phi(\mathbf{x}))] = \frac{\beta^2}{2}\phi^2(\mathbf{x}) - \frac{\beta^4}{12}\phi^4(\mathbf{x}) + ... \qquad (1.98)$$

which, after passing to the continuum limit for $a \to 0$ yields the *Ginzburg-Landau* form of the Hamiltonian

$$\beta H = \int d^d\mathbf{x}\left[\frac{1}{2}(\nabla\phi)^2 + \frac{m^2}{2}\phi^2 + \frac{\lambda}{4!}\phi^4\right]. \qquad (1.99)$$

This theory can easily be treated. We consider the case of a homogeneous order parameter, $\phi = const$. Then

$$\beta V_0(\phi) = \frac{m^2}{2}\phi^2 + \frac{\lambda}{24}\phi^4 \qquad (1.100)$$

is the *mean-field* or *effective potential* of the Ginzburg-Landau theory. Depending on the sign of the quadratic term - with a fourth-order term which has to be strictly positive for thermodynamic stability - the potential displays two shapes which are shown in Figure 1.2. For $m^2 > 0$, the potential has a single minimum at $\phi = 0$; this can be identified with a disordered state. For $m^2 < 0$, two minima appear at values $\phi = \pm\phi_0$. They correspond to a pair of ordered states related to each other by a mirror-symmetry: one has a positive, the other a negative value of the same magnitude. For the ferromagnet these states correspond to states with positive and negative magnetization, and are identical to the solutions obtained from the graphical solution of the saddle-point equation in Figure 1.1.

So far we have not specified the value of m^2; in any case, in the vicinity of the transition at $m = 0$ we know that

$$m^2 \sim \beta_c - \beta \sim |T - T_c| . \qquad (1.101)$$

On the other hand, on dimensional grounds, m must be an inverse length, and we define

$$m \equiv \xi^{-1} \sim |T - T_c|^{1/2} , \qquad (1.102)$$

i.e.,

$$\xi \sim |T - T_c|^{-1/2} . \qquad (1.103)$$

We had encountered ξ before: it is the (spin-spin) correlation length. In eq.(1.103) we have obtained the first *critical exponent* which describes the power-law behaviour of physical quantities in the vicinity of the critical point: the exponent $\nu = 1/2$ is thus the critical exponent of the correlation length.

The information that the approach to a critical point is by a power-law is essential. Power laws are a general characteristic of systems without an intrinsic length scale (an example is a decay-length of correlations), or, in other words, in *scale-invariant* systems. This property of critical systems explains why power laws are so dearly loved by statistical physicists, and why they try to find them in more complex systems as well - a point we will return to in Part III of the book.

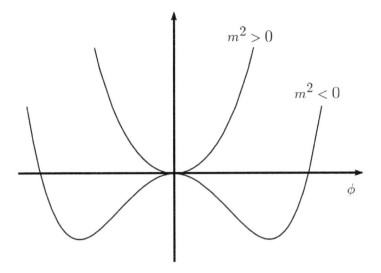

FIGURE I.1.2: Mean-field potential of the Ginzburg-Landau theory, for $T < T_c$ ($m^2 < 0$) and $T > T_c$ ($m^2 > 0$).

The Ginzburg-Landau model is, from a computational point of view, a very nice topic. We therefore suggest a number of exercises and tasks for the readers to try.

Task. Perform the steps explicitly which are needed to pass from the discrete to the continuum version of the Ginzburg-Landau theory, leading to eq.(1.99). Verify that the mass parameter and the coupling constant are given by

$$m^2 = \frac{\beta_c^{-2}}{Ja^2}(\beta_c - \beta), \quad \lambda = \frac{4\beta^2}{J^2\beta_c^4}a^{d-4}. \tag{1.104}$$

Exercise. Perform the mean-field approximation for a homogeneous order parameter at the level of the continuous Ginzburg-Landau theory by adding a field term $-h\phi$. Discuss the phase transition in this model for $h = 0$ at $T = T_c$ and at $T < T_c$ and $h \to 0$.

As it turns out, the phase transition of the Ising model at $T = T_c$, $h = 0$ is a *second-order phase transition*, while for $T < T_c$, $h \to 0$, it is a *first-order phase transition*. This terminology derives from the singular (power-law) behaviour of the first- and second derivatives of the free energy near the transition. At a first-order phase transition, there is a jump in the first derivative of the free energy or, in other words, the free energy has a kink. For a second-order phase transition, the free energy is continuous and differentiable, but at the transition its second derivative displays a discontinuity. We will encounter the distinction between first- and second-order transitions as well in Part II, Chapter 2, in the discussion of the thermal stability of DNA.

Exercise. Draw the effective potential of the Ising model in Ginzburg-Landau theory with and without field. Draw the shape of the potential near the first and second order transition to illustrate the above discussion.

Exercise. Calculate the dependence of the *susceptibility* on the distance of the critical point, $|T - T_c|$, i.e.

$$\chi = \partial_h \phi_0|_{h=0} . \tag{1.105}$$

Deduce the critical exponent of the susceptibility.

Exercise. Consider the phase transition from the disordered to the ordered state at $h = 0$ and at a finite value of h. Calculate the *specific heat*

$$C \equiv -T \partial_T F . \tag{1.106}$$

What do you notice? Determine the critical exponent.

Task. Are there also solutions to the Ginzburg-Landau equations which depend on space, e.g., one-dimensional profiles $\phi(x)$? What is their physical interpretation?

Phenomenological scaling. Having found the first examples of critical exponents, the next obvious question to ask is whether there are any relations between them. In order to answer this question we start from the homogeneous Ginzburg-Landau equation with field and express it in the form

$$m^2 \phi + u \phi^3 = h . \tag{1.107}$$

The solutions to this equation describe a family of solutions $\phi = \phi(m^2, h)$. We have discussed before what happens if $h = 0$; if we approach the critical point at $T = T_c$, hence at $m^2 = 0$, we find the power-law dependence

$$\phi \sim h^{1/3} , \tag{1.108}$$

which yields another critical exponent.

Is there a way to combine the two control parameters m^2 and h? Let us introduce a scaled field variable

$$\psi = \phi h^{-1/3} . \tag{1.109}$$

Going back to eq.(1.107) we find

$$m^2 h^{-2/3} \psi + u \psi^3 = 1 \,. \tag{1.110}$$

Thus by redefining $x \equiv m^2 h^{-2/3}$ we obtain $x\psi + u\psi^3 = 1$ and the whole dependence on m^2 and h is now through a parameter combination. The general solution of the Ginzburg-Landau equation thus will have the *scaling form*

$$\phi(m^2, h) = h^{1/3} \psi(m^2 h^{-2/3}) \,. \tag{1.111}$$

This equation states that the function $\phi h^{-1/3}$ has a *universal form*, and all curves parametrized by x will precisely collapse on this one universal curve.

What is the expression of this function? For the cubic polynomial, one can calculate it explicitly (*Task!*), but in most cases this is not feasible. It is more instructive to see the properties of this function in certain limits. We know that at $x = 0$

$$\psi(0) \sim u^{-1/3} \,. \tag{1.112}$$

For $x \to \infty$, the cubic term in eq.(1.110) can be dropped and we find in that limit

$$\psi(x) \approx \frac{1}{x} \,, \tag{1.113}$$

hence

$$\phi \approx h^{1/3} (m^2 h^{-2/3})^{-1} = \frac{h}{m^2} \sim \frac{h}{T} \tag{1.114}$$

which is the *high-temperature* or *weak-field limit*; the final result is called *Curie's law* of paramagnetism in the context of magnetism.

For the opposite limit, $x \to -\infty$, we have $|x\psi| \gg 1$ and $|u\psi^3| \gg 1$ so that

$$\psi \approx \sqrt{|x|/u} \,. \tag{1.115}$$

This we also have seen before, it is the *ordered phase* with $\phi \sim |m^2|^{1/2}$.

The idea to combine the control parameters into a single scaling variable has proved extremely fruitful for the theory of phase transitions. In the mid-sixties of the last century B. WIDOM proposed that quite generally the free energy per unit volume can be written in the scaling form

$$f(m^2, h) = |m^2|^{2-\alpha} f_s(h(m^2)^\Delta) \,, \tag{1.116}$$

where α is the *specific heat exponent*, and Δ the *gap exponent*. From this hypothesis, a whole sequence of critical exponent relations can be deduced that are now known to hold independent of spatial dimension (with some notable exceptions).

Task. From the above considerations of phenomenological scaling for the Ginzburg-Landau theory, deduce the relation between the critical exponents β, α, and Δ. Here, β is *not* to be confused with the abbreviation of $(k_B T)^{-1}$; it is the common notation for the exponent characterizing the temperature dependence of the order parameter via $m \sim |T - T_c|^\beta$.

Beyond Ginzburg-Landau theory. Ginzburg-Landau theory has permitted us to very easily find the equilibrium states ϕ_0 of the system by an almost trivial analytic calculation, the minimization of the effective potential $V_0(\phi)$, eq.(1.100). Further, we have found the critical exponents. But we still have no clue about the validity of this approach. In contrast to our calculation of the partition function of the one-dimensional Ising model, spatial dimensions do nowhere appear explicitly, and the basic conflict between the one-dimensional result (no phase transition at $T > 0$) and the Ginzburg-Landau result (a phase transition at $T = T_c \neq 0$) persists. In order to finally resolve that conflict we now have to look for the effects of fluctuations around the mean-field solution. We expect that they will modify the behaviour we have obtained.

In order to compute the fluctuations around the mean-field solution we put $\phi = \phi_0(x) + \delta\phi$ and expand the Ginzburg-Landau Hamiltonian in $\delta\phi$. This leads to a partition function

$$Z = \exp[-\beta(H_0 + H_1)] \qquad (1.117)$$

where βH_0 is the GL-value, and βH_1 is determined by a Gaussian integral over the $\delta\phi$-fluctuations which can be carried out exactly. It reads

$$\beta H_1 = \frac{1}{2} \int d^d\mathbf{x} \ln\left[-\nabla_\mathbf{x}^2 + m^2 + \frac{1}{2}\lambda\phi_0^2(\mathbf{x})\right] \delta^d(\mathbf{x} - \mathbf{y})|_{\mathbf{x}=\mathbf{y}} \qquad (1.118)$$

With the Fourier representation of the δ−function[18] we rewrite this as

$$\beta H_1 = \int d^d\mathbf{x} \int \frac{d^d\mathbf{k}}{(2\pi)^d} e^{-i\mathbf{k}\cdot\mathbf{x}} \ln\left[-\nabla_x^2 + m^2 + \frac{1}{2}\lambda\phi_0^2(x)\right] e^{i\mathbf{k}\cdot\mathbf{x}} \qquad (1.120)$$

---

[18]The Fourier representation of a function we use in the following is given by the integral

$$\hat{f}(\mathbf{x}) = \int \frac{d^d\mathbf{k}}{(2\pi)^d} e^{i\mathbf{k}\cdot\mathbf{x}} f(\mathbf{k}) . \qquad (1.119)$$

Note that we frequently drop the ⌢ and distinguish between a function and its Fourier transform by its argument only.

where the term $\sim e^{-i\mathbf{k}\cdot\mathbf{y}}$ was first pulled through to the left and then put to $\mathbf{y} = \mathbf{x}$. If $\phi_0 = const.$ we find

$$\beta H_1 = \int \frac{d^d\mathbf{k}}{(2\pi)^d} \ln\left[\mathbf{k}^2 + m^2 + \frac{1}{2}\lambda\phi_0^2\right]. \tag{1.121}$$

Eq.(1.121) is the main result of this paragraph, and its consequences will now be discussed. Obviously, this integral is not well-behaved for large $\mathbf{k}$-values, indicating the breakdown of the continuum theory for small spatial scales.[19] We can render the integral finite by introducing a cut-off Λ, but already here it becomes evident that spatial dimension now enters in the calculation. In $d = 3$ we obtain

$$\beta H_1 = \frac{\lambda\phi_0^2\Lambda}{4\pi^2} - \frac{1}{6\pi}\left(m^2 + \frac{1}{2}\lambda\phi_0^2\right)^{3/2} + \mathcal{O}(\Lambda^\alpha) \tag{1.122}$$

where higher-order terms depending on Λ are summed up in the last term; further, terms which do not contain a dependence on the field have been dropped.

The expression (1.122) can be understood as an additional contribution to the effective potential in mean-field which we recall had the form

$$\beta V_0 = \frac{1}{2}m^2\phi_0^2 + \frac{\lambda}{4!}\phi_0^4. \tag{1.123}$$

Expanding the root in βH_1 up to fourth order in ϕ_0 we obtain

$$\beta H_1 = -\frac{m^3}{6\pi}m^3 + \frac{\lambda\phi_0^2\Lambda}{4\pi^2} - \frac{\lambda m\phi_0^2}{8\pi} - \frac{1}{64\pi}\frac{\lambda^2}{m}\phi_0^4, \tag{1.124}$$

where the first term can be ignored since it does not depend on ϕ_0. We can now restore the original form of the GL-potential by the redefinitions

$$m_{eff}^2 = m^2 - \frac{\lambda\Lambda}{2\pi^2} + \frac{\lambda m}{4\pi} \tag{1.125}$$

and

$$\lambda_{eff} = \lambda + \frac{3}{8\pi}\frac{\lambda^2}{m}. \tag{1.126}$$

---

[19]This phenomenon is called a *UV-divergence* in field theory since it occurs at large wavevectors, hence 'high energies'. We do not follow this point here in all its consequences, since this goes far beyond what is attempted here. The limited ambition here is to illustrate that a continuum theory has to be considered carefully: it is potentially dangerous to extrapolate its results down to microscopic scales. This may not show up in a purely mean-field approach, but if one wants to go beyond to include fluctuation effects, surprises can happen.

We stress again that, although the effective potential in $d = 3$ has the same form as the effective potential in mean-field, the new result does depend both on spatial dimension (the calculation is only valid in $d = 3$) and on the cutoff Λ we used in calculating the integral.

The vicinity of the transition: $m^2 \to 0$. We are now ready to take the final step and look at the effect of fluctuations at the critical point, or, more precisely, in the vicinity of the phase transition. In order to do this we have to look at the correlation function of the field $\phi(\mathbf{x})$ in the limit $m^2 \to 0$, since this will now give us additional information beyond the value of the order parameter itself, which we can compute from the fluctuation-modified effective potential.

Calling the correlation function of the field[20] $G(\mathbf{x})$, it fulfills the differential equation

$$(-\nabla_{\mathbf{x}}^2 + m^2)G(\mathbf{x}) = \delta^d(\mathbf{x}), \tag{1.127}$$

and, consequently, its Fourier transform is given by

$$G(\mathbf{k}) = \frac{1}{\mathbf{k}^2 + m^2}. \tag{1.128}$$

with, as before, $m \equiv \xi^{-1} \sim |T - T_c|^{1/2}$.

Now let's transform back to real space. We find

$$G(r) = \int \frac{d^d\mathbf{k}}{(2\pi)^d} \frac{e^{i\mathbf{k}\cdot\mathbf{x}}}{\mathbf{k}^2 + \xi^{-2}} = \xi^{2-d} \int \frac{d^d\mathbf{q}}{(2\pi)^d} \frac{e^{i\mathbf{q}\cdot\mathbf{x}/\xi}}{\mathbf{q}^2 + 1}, \tag{1.129}$$

where $\mathbf{q} = \xi\mathbf{k}$. The integrand has two asymptotic limits for $|x| = r$:

$$G(r) \sim \begin{cases} \frac{e^{-r/\xi}}{r^{(d-1)/2}}, & r \gg \xi \\ \\ \frac{1}{r^{d-2}}, & r \ll \xi. \end{cases} \tag{1.130}$$

Note that the first expression indeed reduces to our previous result, eq.(1.75), for $d = 1$. This means that, within Ginzburg-Landau theory, although there is a mean-field transition, fluctuations will destroy this transition and the original (exact) calculation is supported.

---

[20] We made use of translational invariance of the system by shifting one of the arguments to $\mathbf{x} = \mathbf{0}$. In a translationally invariant system, the correlation function depends on the distance of field at the two selected points in space - we saw this already before in our computation of the correlation function in the one-dimensional spin chain.

The existence of a second regime in which correlations decay algebraically signals the presence of a true ordered phase at a finite temperature, i.e., for $0 < T < T_c$, and we see that this is certainly possible for $d > 2$.

With this information we can now, finally, estimate the range of validity of the Ginzburg-Landau theory. Considering length scales on the order of the correlation length, we have

$$G(\xi) \sim \xi^{2-d} \sim |T - T_c|^{\frac{d-2}{2}}. \tag{1.131}$$

If we look at the ratio of G and the square of the mean-field value of the order parameter,

$$\frac{G(\xi)}{\phi_0^2} \sim |T - T_c|^{\frac{d-4}{2}}, \tag{1.132}$$

we see that, in the vicinity of the transition for $T \to T_c$, fluctuations grow indefinitely for $d < 4$. The dimension $d_u = 4$ is hence considered as an *upper critical dimension* above which mean-field theory becomes exact with respect to the critical exponents, i.e., fluctuations are negligible.[21] The dimension $d_l = 1$ is likewise a *lower critical dimension* at which the phase transition is destroyed by fluctuations.

This concludes our discussion of the Ising model. We leave this model at a point when, from the point of view of statistical physics, things become really interesting: how can we mathematically describe the transition in the range of dimensions $1 < d < 4$? In $d = 2$, the Ising model can be solved exactly by a transfer matrix approach (L. ONSAGER, 1944); in $d = 3$ the problem of the computation of the partition function was shown to be NP-complete (S. IsTRAIL, 2000). It is here where renormalization-group methods need to be used, which go beyond what we attempt here. For those who are interested to get a systematic view of the theory of critical phenomena, some corresponding literature is listed at the end of this Chapter.

Before moving on to our second exemplary system, the model of an elastic polymer, it is nevertheless important to put the calculations we went through into context. What we did looked very much like hard-core statistical physics, and we seem to have lost track of biology. So what is this all good for, in the context of computational biology?

The answer that can be given to this question has three aspects. First, as we will see in this book, there are phase transitions in biological systems. The methodology we have developed here is thus needed to understand these

---

[21]In a more detailed calculation one can see that right at the upper critical dimensions, *logarithmic corrections* arise.

transitions. We will see a particular example in Part II of the book when we study the thermal denaturation of DNA. Even before, in Chapter 2 of Part I, we will see that the concept of a phase transition carries over also to nonequilibrium situations. The alignment of DNA or protein sequences can be understood as a nonequilibrium phase transition.

Second, we will see in what follows that even though biological systems do not always have phase transitions, they may nevertheless exhibit scaling behaviour in some of their characteristic quantities. Scaling is thus a useful method of analysis, although one should not be carried away to overinterpret its outcome. We will touch on this aspect in Part II of the book, e.g., in the context of the analysis of microarray data, and in Part III when we discuss global features of biological networks.

And finally, a phase transition is the prime example of a *cooperative phenomenon*. Cooperative phenomena arise in biology in various different contexts, and we will encounter them here, too. We will see a direct biological application of the Ising model to the cooperative behaviour of receptors in the membrane of a bacterium.

We are now ready to turn to our second illustrative example, a model for an elastic (bio-)polymer.

Modelling biopolymer elasticity. As a second application example for the methods of statistical mechanics we introduce a model description for the elastic behaviour of biomolecules. Molecular elasticity is entropic in origin. This merits an explanation.

Let's begin with a simplistic version of a polymer chain model which is called the *freely jointed chain*. We imagine the polymer can be abstracted as a chain of N freely jointed links, as illustrated in Figure 1.3. The spatial configuration of the chain resembles the trajectory of a *random walk*.

If we now pull at the ends of our caricature polymer with a force **F**, we will straighten out the chain. What happens if we then let go? Imagining we have realized the chain as made from paperclips, it would simply stay straight, as there is no internal or external mechanism that would pull the segment back into the initial random configuration. Something is missing, this idea is wrong.

A biomolecule in solution is subjected to thermal energy, $k_B T$. The positional fluctuations of the surrounding molecules hit the chain and will thereby gently randomize the configuration. Straightening out the chain against these thermally fluctuating forces consequently reduces the entropy of the chain, i.e., the number of conformations it can adopt. Work has to be performed, and the relation between force and extension of a molecule in solution results

FIGURE I.1.3: The simplistic polymer model: a chain of linear segments which can freely move about their links.

thus from an *entropic elasticity*.

Pulling DNA. Based on this concept, various experimental techniques have been developed to exert forces on molecules in solution. The basic principle is that the molecule has to be fixed at one end, while the other can be manipulated by mechanical force. A typical setup is that a DNA molecule is chemically modified at its both extremities; it is then rigidly fixed at one end to a surface (a capillary), and to a superparamagnetic bead at the other end. The application of a magnetic field allows to pull on the bead and hence the DNA molecule. Another possible setup is by trapping a bead in an optical trap; such a setup is indicated schematically in Figure 1.4. Questions of biological or biophysical interest that have been addressed with this approach are listed in the additional notes at the end of this chapter; here we only want to understand how to describe the elastic properties.

The WLC-model. Following the suggested discrete representation of the polymer as a chain of elements, we first consider the polymer as a linear chain with element length b, such that the chain made of N segments has a total length of $L = bN$. The energy of the chain is given by the Hamiltonian

$$\beta H = -K \sum_i \widehat{\mathbf{t}}_i \cdot \widehat{\mathbf{t}}_{i+1} - \beta F b \sum_i \widehat{\mathbf{t}}_i \cdot \widehat{z} \qquad (1.133)$$

where an applied force F is directed along the z-axis.

In eq.(1.133), K/β is the *stiffness* of the chain, i.e., its resistance to bending. The unit vector $\widehat{\mathbf{t}}_i$ describes the orientation of the segment i. This model is sometimes called the *Kratky-Porod model*, which for $K = 0$ reduces to the

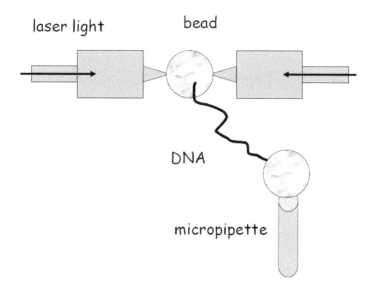

FIGURE I.1.4: A setup for DNA stretching experiments using an optical trap.

freely jointed chain we had pictured before.

In the continuum limit $b \to 0$, where $K \equiv \ell_P/b$ introduces the *persistence length* ℓ_P, we can represent the model as the *Worm-Like Chain (WLC)* model with the energy expression

$$\beta H_{WLC} = \frac{\ell_P}{2} \int_0^L ds (\partial_s \widehat{\mathbf{t}})^2 - f \int_0^L ds\, \widehat{\mathbf{t}} \cdot \widehat{z} \qquad (1.134)$$

where the pulling force per unit length is $F = k_B T f$. In this formulation, s is the curvilinear coordinate along the chain, and $\widehat{\mathbf{t}}$ has turned from the orientational vector of a chain element into the tangential vector along the continuous curve which makes out the chain.

As noted, the first term in the Hamiltonian is the *bending energy*. Its coefficient, the persistence length, is a length scale measure for the correlations of the tangent vectors along the chain, or, in other words, for the propagation of changes in the chain conformation along the chain.[22]

---

[22]There are several other commonly used lengths that characterize polymers; a prominent one is the *Kuhn length* with $\ell_K = 2\ell_p$, which is well-defined provided $L \gg \ell_K$.

As stated above, the quantity we are interested in from an experimental point of view is the force-extension relation. For the WLC-model, the exact computation of the partition function is equivalent to solving the problem of a quantum rotator subjected to a polarizing field. Here we give a simplified calculation valid for the case when the forces pulling at the molecule are sufficiently strong and limit the chain excursions transverse to the pulling direction (J. F. MARKO and E. SIGGIA, 1995).

Since $|\widehat{\mathbf{t}}| = 1$, we can take the transverse components $(t_x, t_y) \equiv \widehat{\mathbf{t}}_\perp$ as independent variables, and obtain for the z-component

$$t_z = 1 - \widehat{\mathbf{t}}_\perp^2/2 + O(\widehat{\mathbf{t}}_\perp^4). \tag{1.135}$$

We can then express the energy in a Gaussian approximation as

$$\beta H_{WLC} = \frac{1}{2}\int_0^L ds[\ell_p(\partial_s\widehat{\mathbf{t}}_\perp)^2 + f\widehat{\mathbf{t}}_\perp^2] - fL. \tag{1.136}$$

Introducing the Fourier modes

$$\tilde{\mathbf{t}}_\perp(q) = \int ds\, e^{iqs}\,\widehat{\mathbf{t}}_\perp(s) \tag{1.137}$$

we obtain

$$\beta H_{WLC} = \frac{1}{2}\int \frac{dq}{2\pi}[\ell_p q^2 + f]\tilde{\mathbf{t}}_\perp^2 - fL. \tag{1.138}$$

The average of the transverse components along the chain s follows from equipartition

$$\langle\widehat{\mathbf{t}}_\perp^2\rangle = \int \frac{dq}{2\pi}\langle|\tilde{\mathbf{t}}_\perp(q)|^2\rangle = 2\int \frac{dq}{2\pi}\left[\frac{1}{\ell_p q^2 + f}\right] = \frac{1}{\sqrt{f\ell_P}} \tag{1.139}$$

where the factor of two accounts for the two components of $\widehat{\mathbf{t}}_\perp$. The extension of the chain is then obtained as

$$\frac{z}{L} = \widehat{\mathbf{t}}\cdot\widehat{\mathbf{z}} = 1 - \frac{\langle\widehat{\mathbf{t}}_\perp^2\rangle}{2} = 1 - \frac{\sqrt{f\ell_P}}{2} \tag{1.140}$$

which behaves as a square-root in f. The approximate formula

$$f = \frac{1}{\ell_P}\left[\frac{z}{L} + \frac{1}{4(1 - z/L)^2} - \frac{1}{4}\right] \tag{1.141}$$

describes the experimental behaviour rather well, see Figure 1.5.

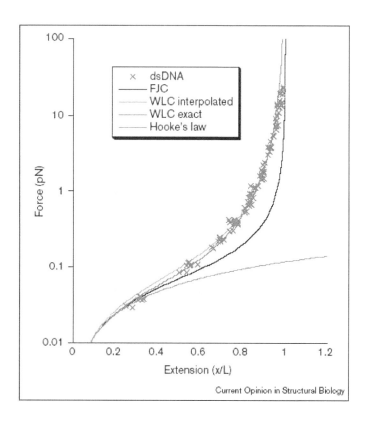

FIGURE I.1.5: Force-extension curve for a single-molecule DNA. (Reprinted from C. BUSTAMANTE et al., 2000, with permission from Elsevier.)

Additional Notes

Equilibrium statistical mechanics is a well-developed theory on which many books have been and are being published, written to every taste of mathematical rigor. Four suggestions are the books by K. HUANG, M. PLISCHKE and B. BIRGERSEN, F. SCHWABL and L. E. REICHL. Books focussing on the renormalization group methods of statistical physics are, again as examples, the volumes by L. P. KADANOFF and N. GOLDENFELD.

A highly recommendable introduction into biophysics, touching on many aspects of statistical physics, has been written by P. NELSON.

Experiments on single molecule manipulations of nucleic acids have started in the early nineties of the last century; basic references on DNA elasticity are T. STRICK et al, 1996 and 1998. Reviews on the basic physics and the experimental techniques are C. BUSTAMANTE et al., 2000, and U. BOCKELMANN, 2004.

The advantage of the single-molecule technique is that the mechanical force exerted on the molecule can be made to interfere with enzymatic reactions on the chains, allowing for a mechanical probing of biochemical processes. Examples from the catalogue of achievements are:

- Unwinding of promoter regions in transcription initiation: A. REVJAKIN et al., 2004.

- Pausing of RNA polymerase during transcription: R. J. DAVENPORT et al., 2000 and N. R. FORDE et al., 2002.

- Base-pair stepping of RNA polymerase: E. A. ABBONDANZIERI et al., 2005.

- Torque-induced structural transitions in DNA, of relevance for enzymatic reactions: Z. BRYANT et al., 2003.

- DNA unzipping: C. DANILOWICZ et al., 2003.

Theoretical papers on DNA elasticity are the classic paper on the wormlike chain model by J. F. MARKO and E. D. SIGGIA, 1995. Extensions to include twist were published by C. BOUCHIAT and M. MÉZARD, 1997, and J. D. MOROZ and P. NELSON, 1997.

References

E. A. Abbondanzieri, W. J. Greenleaf, J. W. Shaevitz, R. Landick and S. M. Block, *Direct observation of base-pair stepping by RNA polymerase*, Nature **438**, 460-465 (2005)

U. Bockelmann, *Single-molecule manipulation of nucleic acids*, Curr. Op. Struct. Biol. **14**, 368-373 (2004)

C. Bouchiat and M. Mézard, *Elasticity Model of a Supercoiled DNA Molecule*, Phys. Rev. Lett. **80**, 1556-1559 (1998)

Z. Bryant, M. D. Stone, J. Gore, S. B. Smith, N. R. Cozzarelli and C. Bustamante, *Structural transitions and elasticity from torque measurements on DNA*, Nature **424**, 338-341 (2003)

C. Bustamante, S. B. Smith, J. Liphardt and D. Smith, *Single-molecule studies of DNA mechanics*, Curr. Op. Struct. Biol. **10**, 279-285 (2000)

C. Danilowicz, V. W. Coljee, C. Bouzigues, D. K. Lubensky, D. R. Nelson and M. Prentiss, *DNA unzipped under a constant force exhibits multiple metastable intermediates*, Proc. Natl. Acad. Sci. USA **100**, 1694-1699 (2003)

R. J. Davenport, G. J. L. Wuite, R. Landick and C. Bustamante, *Single-Molecule Study of Transcriptional Pausing and Arrest by E. coli RNA Polymerase*, Science **287**, 2497-2500 (2000)

N. R. Forde, D. Izhaky, G. R. Woodcock, G. J. L. Wuite and C. Bustamante, *Using mechanical force to probe the mechanism of pausing and arrest during continuous elongation by Escherichia coli RNA polymerase*, Proc. Natl. Acad. Sci. USA **99**, 11682-11687 (2002)

N. Goldenfeld, *Lectures on Phase Transitions and the Renormalization Group*, Westview Press (1992)

K. Huang, *Introduction to Statistical Physics*, CRC Press (2001)

S. Istrail, *Statistical Mechanics, Three-Dimensionality and NP-Completeness: I. Universality of Intractability of the Partition Functions of the Ising Model Across Non-Planar Lattices*, Proc. 32nd ACM Symposium on the Theory of Computing (STOC00), ACM Press, 87-96 (2000)

E. T. Jaynes, *Information Theory and Statistical Mechanics I.*, Phys. Rev. **106**, 620-630 (1957)

E. T. Jaynes, *Information Theory and Statistical Mechanics II.*, Phys. Rev. **108**, 171-190 (1957)

L. P. Kadanoff, *Statistical Physics. Statics, Dynamics and Renormalization*, World Scientific (2000)

J. Majewski, H. Li and J. Ott, *The Ising Model in Physics and Statistical Genetics*, Am. J. Hum. Genet. **69**, 853-862 (2001)

J. F. Marko and E. D. Siggia, *Stretching DNA*, Macromolecules **28**, 8759-8770 (1995)

J. D. Moroz and P. Nelson, *Torsional directed walks, entropic elasticity, and DNA twist stiffness*, Proc. Natl. Acad. Sci. USA **94**, 14418-14422 (1997)

P. Nelson, *Biological Physics. Energy, Information, Life*, Freeman (2004)

L. Onsager, *Crystal statistics. I. A two-dimensional model with an order-disorder transition*, Phys. Rev. **65**, 117-149 (1944)

M. Plischke and B. Birgersen, *Equilibrium Statistical Physics*, Scientific Publishers 2nd ed. (1994)

L. E. Reichl, *A Modern Course in Statistical Physics*, Wiley-Interscience 2nd ed. (1998)

A. Revjakin, R. H. Ebright and T. R. Strick, *Promoter unwinding and promoter clearance by RNA polymerase: Detection by single-molecule DNA nanomanipulation*, Proc. Natl. Acad. Sci. USA **101**, 4776-4780 (2004)

F. Schwabl, *Statistical Mechanics*, Springer (2002)

T. R. Strick, J.-F. Allemand, D. Bensimon, A. Bensimon and V. Croquette, *The Elasticity of a Single Supercoiled DNA Molecule*, Science **271**, 1835-1837 (1996)

T. R. Strick, J.-F. Allemand, V. Croquette and D. Bensimon, *Physical Approaches to the Study of DNA*, J. Stat. Phys. **93**, 647-672 (1998)

B. Widom, *Equation of State in the Neighborhood of the Critical Point*, J. Chem. Phys. **43**, 3898-3905 (1965)

Chapter 2

Nonequilibrium Statistical Mechanics

Nonequilibrium statistical mechanics does not yet have such powerful general concepts and tools as equilibrium statistical mechanics. Attempts to develop such concepts for physical processes which occur arbitrarily far from a thermal equilibrium situation have so far not led to the desired success.[1]

Since we still lack these general principles which govern the time-dependent or stationary distributions of our physical quantities of interest, we instead have to directly address the properties of the stochastic processes themselves and try to find methods to treat them, at least with some reasonable simplifications.

In the most general case we can address, we will be able to compute the time evolution of suitably defined probability distributions characterizing the state of a physical system, or its transition from one state to another. Consequently, our symbol of choice in this Part is P rather than Z.

What we will also see, however, is that we sometimes will be able to relate nonequilibrium processes to equilibrium processes. We will find expressions, the so-called *(dissipation-)fluctuation theorems*, which establish such a relation. Most often, however, these relations require for their validity to be not too far away from a thermal equilibrium. There is one notable exception we will address as well, a result obtained some years ago by C. JARZYNSKI.

We begin this Chapter with the formal bits and pieces of a general description of stochastic processes and their time evolution.[2]

---

[1]Of course, there is an extended body of work on irreversible processes. Nevertheless, I think there is a general consensus that a theory of noneqilibrium statistical physics of the same level of generality as that of equilibrium statistical physics has not yet been established.

[2]The presentation, in particular at the beginning, follows the book by N. VAN KAMPEN, 1992, which can be seen as a standard reference on most of the topics discussed in this Chapter.

2.1   Stochastic processes

The first section of this Chapter lists a number of definitions for later use.

We call X be a *stochastic variable* with a particular value x; f is considered a mapping from X at time t. We call

$$Y_X(t) = f(X, t) \tag{2.1}$$

a *random function*, and $Y_x(t) = f(x, t)$ a *sample function* or a *realization* of the stochastic process; as for equilibrium states, we can speak of the corresponding ensemble in a natural way. For the stochastic process, we define an *ensemble average* by

$$\langle Y(t) \rangle \equiv \int dx\, Y_x(t)\, P(x). \tag{2.2}$$

The higher moments of the distribution are defined in an analogous way. The probability for $Y_x(t)$ to take the value y at time t is given by

$$P(y, t) = \int dx P(x) \delta(y - Y_x(t)). \tag{2.3}$$

The mean and the moments are, as in thermal equilibrium, quantities that allow to quantify the probability distributions. For time-dependent processes, another quantity is of interest:

The autocorrelation function. The *autocorrelation function* of the stochastic process is defined by

$$A(t_1, t_2) \equiv \langle Y(t_1) Y(t_2) \rangle - \langle Y(t_1) \rangle \langle Y(t_2) \rangle. \tag{2.4}$$

For $t_1 = t_2 = t$, it reduces to the time-dependent variance, $\sigma^2(t)$.

Joint probability density. We define the *joint probability density*

$$P(y_1, t_1; ...; y_n; t_n) \tag{2.5}$$

which states that $Y_x(t)$ has the value y_1 at t_1, ..., y_n at t_n.

Exercise. Write down the mathematical expression for P.

Conditional probability. The *conditional probability* is the probability density for $Y_x(t)$ to take on the value y_2 at t_2 if it was y_1 at t_1, with the normalized density

$$\int dy_2 P(y_2, t_2 | y_1, t_1) = 1. \tag{2.6}$$

Markov process. A stochastic process has the *Markov property* if for any set of ordered timesteps $t_1 < t_2 < ... < t_n$ the conditional probability satisfies

$$P(y_n, t_n | y_1, t_1, ..., y_{n-1}, t_{n-1}) = P(y_n, t_n | y_{n-1}, t_{n-1}). \tag{2.7}$$

P is then a *transition probability*. It only depends on the two states involved, the one that is left and the one that is reached. A Markov process is uniquely determined from the knowledge of $P(y_1, t_1)$ and $P(y_2, t_2 | y_1, t_1)$.

Exercise. Convince yourself of the correctness of the last statement for a process involving three steps, $t_1 < t_2 < t_3$.

The Chapman-Kolmogorov equation. From the transition probabilities of a Markov process we have the expression

$$P(y_3, t_3 | y_1, t_1) = \int dy_2 \, P(y_3, t_3 | y_2, t_2) P(y_2, t_2 | y_1, t_1) \tag{2.8}$$

in which time-ordering is essential.

Examples. a) For $-\infty < y < \infty$, the Chapman-Kolmogorov equation is solved by (for $t_2 > t_1$)

$$P(y_2, t_2 | y_1, t_1) = \frac{1}{\sqrt{2\pi(t_2 - t_1)}} \exp\left[-\frac{(y_2 - y_1)^2}{2(t_2 - t_1)}\right] \tag{2.9}$$

If $P(y, 0) = \delta(y)$, this Markov process is called the *Wiener* or *Wiener-Lévy process*.

b) If $Y_x(t)$ takes on only positive integer values $n = 0, 1, 2, ...$ for $t \geq 0$, eq.(2.8) is obeyed by the *Poisson process*

$$P(n_2, t_2 | n_1, t_1) = \frac{(t_2 - t_1)^{n_2 - n_1}}{(n_2 - n_1)!} e^{-(t_2 - t_1)} \tag{2.10}$$

and

$$P(n, 0) = \delta_{n,0}. \tag{2.11}$$

Task. Show that the probability density of the Wiener process fulfills the *diffusion equation*

$$\partial_t P(x, t) = D \, \partial_x^2 P(x, t) \tag{2.12}$$

with a diffusion constant $D = 1/2$.

Stationary Markov processes. A stochastic process is called *stationary* if all joint probability densities depend only on time-differences τ

$$P(y_1, t_1; ....; y_n, t_n) = P(y_1, t_1 + \tau; ...; y_n, t_n + \tau). \tag{2.13}$$

In this case we write for the conditional probabilities

$$P(y_2, t_2|y_1, t_1) = T_\tau(y_2|y_1) \tag{2.14}$$

and we will also suppress the index τ whenever no ambiguity can arise. With this notation, the Chapman-Kolmogorov equation is rewritten as

$$T(y_3|y_1) = \int dy_2 T(y_3|y_2) T(y_2|y_1). \tag{2.15}$$

Task. Show that the autocorrelation function of a stationary Markov process with zero mean is given by

$$A(\tau) = \int \int dy_1 \, dy_2 \, y_1 \, y_2 \, T_\tau(y_2|y_1) \, P(y_1) \tag{2.16}$$

for $\tau \geq 0$.

Example. The standard example of a stationary Markov process is the *Ornstein-Uhlenbeck process*, for which

$$P(y_1) = \frac{1}{\sqrt{2\pi}} \exp\left(-\frac{y_1^2}{2}\right) \tag{2.17}$$

and

$$T(y_2|y_1) = \frac{1}{\sqrt{2\pi(1 - e^{-2\tau})}} \exp\left[-\frac{(y_2 - y_1 e^{-\tau})^2}{2(1 - e^{-2\tau})}\right] \tag{2.18}$$

where $\tau \equiv t_2 - t_1$.

Exercise. Show that the Ornstein-Uhlenbeck process fulfills $\langle Y \rangle = 0$ and $A(\tau) = e^{-\tau}$.

2.2   The master equation

With this minimal list of definitions we have set the stage for the description of stochastic processes within the context of statistical physics, and are ready to move on.[3] The next and very important step is to make practical use of the Chapman-Kolmogorov equation for more complex situations than we have discussed so far in the few examples. The corresponding mathematical tool is the *master equation*.

The master equation is a limiting expression one can obtain from the Chapman-Kolmogorov equation if one lets the time interval tend to zero in a controlled way such that one can pass over to a differential equation for the transition probabilities. Let's do this.

Suppose we write down the transition probability T for the case of small time intervals τ' in the following way

$$T_{\tau'}(y_2|y_1) = (1 - a_0(y_1)\tau')\delta(y_2 - y_1) + \tau'W(y_2|y_1) + o(\tau'^2). \qquad (2.19)$$

This is an expansion to linear order in τ', in which $W(y_2|y_1)$ is defined as the transition probability per unit time from y_1 to y_2, hence $W(y_2|y_1) \geq 0$.

The second term in eq.(2.19) is thus clear, but we have still to say something about the first term. In this term, the coefficient in front of the δ-function takes into account that *no transition occurs* during the time interval τ', and hence we must have for the coefficient a_0

$$a_0(y_1) = \int dy_2 W(y_2|y_1). \qquad (2.20)$$

Finally, the last term in eq.(2.19) denotes terms of higher order in τ' which we neglect in the following.

We insert this expression into the Chapman-Kolmogorov equation and do some rearrangement of terms, with the result

$$\frac{T_{\tau+\tau'}(y_3|y_1) - T_\tau(y_3|y_1)}{\tau'} = -a_0(y_3) T_\tau(y_3|y_1) + \int dy_2 W(y_3|y_2)T_\tau(y_2|y_1)$$
$$(2.21)$$

which in the limit $\tau' \to 0$ yields the differential equation

$$\partial_\tau T_\tau(y_3|y_1) = \int dy_2 \left[W(y_3|y_2)T_\tau(y_2|y_1) - W(y_2|y_3)T_\tau(y_3|y_1)\right]. \qquad (2.22)$$

This is the master equation in its continuous form. Usually it is rewritten in

---

[3]A reader who wishes to get more is asked to consult VAN KAMPEN's book; see the list of references.

a more intuitive form as

$$\partial_t P(y,t) = \int dy' \left[W(y|y')P(y',t) - W(y'|y)P(y,t) \right] . \qquad (2.23)$$

This equation must be solved for $t \geq t_1$ given an initial condition $P(y,t_1) = \delta(t - t_1)$. Note that the equation should not be misinterpreted as an equation for a single-time distribution, which is a frequent error due to the abusive notation with P; the Chapman-Kolmogorov equation makes it clear that we are considering transition probabilities.

Having our tool finally at hand, we can now turn to some illustrative examples. For this, we employ the master equation in a form applicable to discrete states n, and write it as

$$\dot{p}_n(t) = \sum_m \left[w_{nm} p_m(t) - w_{mn} p_n(t) \right] . \qquad (2.24)$$

This equation can also be understood as a gain-loss equation for the probability of states n with $w_{nm} \geq 0$.

Example. As an example of a non-stationary Markov process, we consider *protein degradation*. Within a cell, proteins are continually degraded by a dedicated cell machinery. In our simple model we describe this by a rate γ. The transition probability is given by

$$w_{nm} = \gamma\, m\, \delta_{n,m-1} \qquad (2.25)$$

with $n \neq m$. The master equation of this process reads

$$\dot{p}_n = \gamma(n+1)p_{n+1}(t) - \gamma n p_n(t) \qquad (2.26)$$

which needs to be solved under the initial condition $p_n(0) = \delta_{n,n_0}$, i.e., we assume that at time $t = 0$ there are n_0 proteins present.

The simple way to 'solve' eq.(2.26) is to consider the evolution of the average number of proteins, $\langle n(t) \rangle$. This is done by multiplying the equation by n and summing up

$$\sum_{n=0}^{\infty} n\dot{p}_n = \gamma \sum_{n=0}^{\infty} n(n+1)p_{n+1} - \gamma \sum_{n=0}^{\infty} n^2 p_n$$

$$= \gamma \sum_{n=0}^{\infty} (n-1)n p_n - \gamma \sum_{n=0}^{\infty} n^2 p_n \qquad (2.27)$$

$$= -\gamma \sum_{n=0}^{\infty} n p_n$$

which is nothing but

$$\frac{d}{dt}\langle n(t)\rangle = -\gamma\langle n(t)\rangle \tag{2.28}$$

solved by

$$\langle n(t)\rangle = n_0 e^{-\gamma t} \ . \tag{2.29}$$

This is, of course, not the full solution to the problem - but it nicely shows how the behaviour of the average can be obtained.

In order to really solve the master equation we can make use of the generating function

$$G(s,t) = \sum_{n=0}^{n_0} p_n(t)s^n \tag{2.30}$$

defined for $|s| \leq 1$. Multiplying eq.(2.26) by s^n and summing up, the master equation is transformed into a first-order partial differential equation for G in the variables s, t,

$$\partial_t G(s,t) + \gamma(s-1)\partial_s G(s,t) = 0 \ . \tag{2.31}$$

The form of the equation indicates that a *separation of variables* will be of help. We are thus led to write the following *ansatz*[4] for G

$$G(s,t) = (a(t)(s-1)+b)^{n_0} \ , \tag{2.32}$$

where the function $a(t)$ and the constant b need to be determined. Plugging the ansatz into eq.(2.31), the s-dependence drops out, and the function $a(t)$ is found to fulfill the simple differential equation (compare to eq.(2.28))

$$\frac{\dot{a}(t)}{a(t)} = -\gamma \tag{2.33}$$

with the solution $a(t) = e^{-\gamma t}$. The constant $b = 1$, as follows from the initial condition $G(s,0) = s^{n_0}$. Thus

$$G(s,t) = (1+e^{-\gamma t}(s-1))^{n_0} \ . \tag{2.34}$$

---

[4]Note that according to its definition, G is a polynomial in the variable s of order n_0.

Now we have two expressions for $G(s,t)$, the power series in s and its sum. Since the $p_n(t)$ are the coefficients of the series, we only have to Taylor expand eq.(2.34), which leads to the final result

$$p_n(t) = \binom{n_0}{n} \exp(-\gamma n t)(1 - \exp(-\gamma t))^{n_0 - n} . \tag{2.35}$$

One-step processes. A frequently occurring class of stochastic processes that can be studied with the help of master equations are *one-step processes* in which transitions occur only between neighbouring state labels. Writing the master equation as a matrix equation[5]

$$\dot{p}_n = W_{nm} p_m \tag{2.36}$$

we define the matrix W_{nm} as

$$W_{nm} = f_m \delta_{n,m-1} + b_m \delta_{n,m+1} \tag{2.37}$$

with $n \neq m$. The diagonal element of the matrix is given by

$$W_{nn} = -(f_n + b_n) \tag{2.38}$$

so that the full equation reads as

$$\dot{p}_n = f_{n+1} p_{n+1} + b_{n-1} p_{n-1} - (f_n + b_n) p_n . \tag{2.39}$$

The last equation can be written in a more concise form using the *step operator* $\mathcal{E}$, defined by its action to "move up or down the ladder" of states given by h_n:

$$\mathcal{E} h_n = h_{n+1} , \quad \mathcal{E}^{-1} h_n = h_{n-1} \tag{2.40}$$

so that for eq.(2.39) we have

$$\dot{p}_n = (\mathcal{E} - 1) f_n p_n + (\mathcal{E}^{-1} - 1) b_n p_n . \tag{2.41}$$

Exercise. Solve the master equation (2.39) for the *symmetric random walk* with $f_n = b_n = 1$,

$$\dot{p}_n = p_{n+1} - p_{n-1} - 2 p_n , \quad -\infty < n < \infty \tag{2.42}$$

for the initial data $p_n(0) = \delta_{n,0}$.

---

[5]Which makes the linearity of the equation evident.

2.3 Fluctuation theorems

We have mentioned in the introduction to this Chapter that nonequilibrium statistical mechanics still lacks general concepts when compared to equilibrium statistical mechanics. In this section we will derive a theorem (or, an expression) which describes a key property of nonequilibrium systems.

The fluctuation-dissipation theorem. We derive a general fluctuation theorem, valid for nonequilibrium stationary states (and, a fortiori, near equilibrium states) within the formalism we have introduced (M. LAX, 1960).

Suppose we have I molecular species present with numbers n_i, $1 \leq i \leq I$; we consider the index i as a vector index on particle numbers, and abbreviate $\mathbf{n} = (n_1, ... n_I) = n_i$, which is hence a row vector.

From our previous results we infer that the particle distribution function for a Markov process fulfills

$$P(\mathbf{n}, t + \Delta t) = \int d\mathbf{m} P(\mathbf{n}, t + \Delta t | \mathbf{m}, t) P(\mathbf{m} | t), \qquad (2.43)$$

We now define the following quantities[6]

$$\mathbf{A}(\mathbf{m}) = \frac{1}{\Delta t} \int d\mathbf{n} P(\mathbf{n}, t + \Delta t | \mathbf{m}, t)(\mathbf{n} - \mathbf{m}), \qquad (2.44)$$

which is a vector; we call it a *drift vector*.[7] Further, we define the matrix

$$\mathbf{D}(m) = \frac{1}{2\Delta t} \int d\mathbf{n} P(\mathbf{n}, t + \Delta t | \mathbf{m}, t)(\mathbf{n} - \mathbf{m}) \cdot (\mathbf{n} - \mathbf{m})^T \qquad (2.45)$$

where $(...)^T$ denotes the transpose of the row vector of particle numbers, hence a column-vector. We call $\mathbf{D}$ a *diffusion matrix*.

Supposing that the stationary state is characterized by a particle number vector $\mathbf{n}_0$, we introduce the vector

$$\delta \mathbf{n} \equiv \mathbf{n} - \mathbf{n}_0. \qquad (2.46)$$

We now find (verify this as a *Task*) the equation for the time evolution of the mean-value

$$\frac{d}{dt} \langle \delta \mathbf{n} \rangle = \langle \mathbf{A}(\mathbf{n}(t)) \rangle \qquad (2.47)$$

---

[6] This step assumes that the conditional probabilities are expandable to linear order in Δt.
[7] The names of the objects will become clear later.

and the corresponding equation for the covariance matrix

$$\frac{d}{dt}\langle \delta \mathbf{n}\, \delta \mathbf{n}^T\rangle = 2\langle \mathbf{D}(\mathbf{n})\rangle + \langle \mathbf{A}(\mathbf{n})\delta \mathbf{n}^T\rangle + \langle \delta \mathbf{n}\mathbf{A}^T(\mathbf{n})\rangle\,. \tag{2.48}$$

As a following step we expand around the stationary state via $\mathbf{n} = \mathbf{n}_0 + \delta \mathbf{n}$ and assume for the drift-vector

$$\mathbf{A}(\mathbf{n}) \approx \mathbf{A}(\mathbf{n}_0) - \mathbf{\Lambda}\delta \mathbf{n}\,, \tag{2.49}$$

where we have introduced a matrix $\mathbf{\Lambda}$. In this last equation we choose

$$\mathbf{A}(\mathbf{n}_0) = 0 \tag{2.50}$$

to be consistent with eq.(2.47). Further, we make the simplifying assumption that

$$\mathbf{D}(\mathbf{n}) \approx \mathbf{D}(\mathbf{n}_0) \equiv \mathbf{D}\,. \tag{2.51}$$

We then end up with the following results for the mean

$$\frac{d}{dt}\langle \delta \mathbf{n}\rangle = -\mathbf{\Lambda}\langle \delta \mathbf{n}\rangle \tag{2.52}$$

and the covariance matrix

$$\frac{d}{dt}\langle \delta \mathbf{n}\delta \mathbf{n}^T\rangle = 2\mathbf{D} - \mathbf{\Lambda}\langle \delta \mathbf{n}\delta \mathbf{n}^T\rangle - \langle \delta \mathbf{n}\delta \mathbf{n}^T\rangle \mathbf{\Lambda}^T\,. \tag{2.53}$$

These equations are the main result of this section. We note that in steady-state, a relation between the diffusion matrix $\mathbf{D}$, the fluctuations in particle number $\delta \mathbf{n}$ and the drift matrix $\mathbf{\Lambda}$ is established in the form

$$2\mathbf{D} = \mathbf{\Lambda}\langle \delta \mathbf{n}\delta \mathbf{n}^T\rangle + \langle \delta \mathbf{n}\delta \mathbf{n}^T\rangle \mathbf{\Lambda}^T\,. \tag{2.54}$$

This equation is a general version of the so-called *dissipation-fluctuation theorem*. The name derives from special cases - to which we will come in the following - in which the drift vector corresponds, e.g., to a friction force acting on a particle and hence provides a mechanism for dissipation.

The Jarzynski equality. We now turn to another general result, the so-called Jarzynski equality. This time we start the derivation from the master equation, following U. SEIFERT, 2004.

We consider a situation in which we allow the w_{mn} in the master equation to depend on some tunable parameter λ, i.e., we have

$$w_{mn} = w_{mn}(\lambda)\,. \tag{2.55}$$

For a fixed value of λ, we assume (as before) that the system is in a stationary state p_n^s which obeys the condition of *detailed balance*. This condition is quite important: while obeyed by equilibrium systems, the reverse it not true. The condition of detailed balance suffices for the system to have a stationary state, and hence is an important information to have on a nonequilibrium system. Mathematically, the condition of detailed balance is given by

$$\frac{p_n^s}{p_m^s} = \frac{w_{mn}}{w_{nm}}. \tag{2.56}$$

If we now assume that the parameter λ is turned on in a time-dependent manner, $\lambda = \lambda(\tau)$, we would like to know the probability P to encounter a particular trajectory of the system $n(\tau) \equiv (n_0, n_1, ..., n_k)$ starting in n_0 at time $\tau_0 = 0$, jumping to n_1 after a time-interval τ_1 and so forth, until the final jump from τ_{k-1} to $\tau_k \equiv t$. This probability is given by

$$P[n(\tau), \lambda(\tau)] = p^s(n_0, \lambda(0)) \times \prod_{i=1}^{k-1} \exp\left[-\int_{\tau_i}^{\tau_{i+1}} d\tau \sum_{m \neq n_i} w_{n_0, m}(\lambda(\tau))\right]$$
$$\times w_{n_0, m}(\lambda_{\tau_{i+1}}). \tag{2.57}$$

Likewise, we can study the trajectory $\tilde{n} \equiv n(t-\tau)$ which occurs under reversal of λ, i.e., $\lambda(t - \tau) \equiv \tilde{\lambda}$. This operation allows to write down the probability $P[\tilde{n}(\tau), \tilde{\lambda}(\tau)]$ (*Exercise*).

We can then form the ratio of the two probabilities which is given by

$$e^{-R[n(t)]} \equiv \frac{P[\tilde{n}(\tau), \tilde{\lambda}(\tau)]}{P[n(\tau), \lambda(\tau)]} = \exp\left[-\int_0^t d\tau \epsilon'_{n(\tau)} \dot{\lambda}(\tau)\right] \tag{2.58}$$

where the quantity

$$\epsilon'_n(\lambda) = -\frac{d}{d\lambda} \ln p^s(n, \lambda) \tag{2.59}$$

has been introduced. It can be considered as a formal 'energy level'. This interpretation can be made if one wishes to read the integral in the argument of the exponential function as an analogue of a free energy - we will soon see that this analogy can be made precise.

From these observations we obtain the following two identities

$$1 = \sum_{\tilde{n}(\tau)} P[\tilde{n}(\tau), \tilde{\lambda}(\tau)] = \sum_{\tilde{n}(\tau)} e^{-R[n(\tau)]} P[n(\tau), \lambda(\tau)] \tag{2.60}$$

and

$$1 = \sum_{n(\tau)} e^{-R[n(\tau)]} P[n(\tau), \lambda(\tau)] = \left\langle \exp\left(-\int_0^t d\tau \epsilon'_{n(\tau)} \dot\lambda(\tau)\right)\right\rangle . \qquad (2.61)$$

We interpret the second expression as a *fluctuation theorem*

$$\left\langle \exp\left(-\int_0^t d\tau \epsilon'_{n(\tau)} \dot\lambda(\tau)\right)\right\rangle = 1 , \qquad (2.62)$$

and explain now why, by some illustrative applications.

Illustration of the fluctuation theorem. We first illustrate the result eq.(2.62) by applying it to a simple cyclically working enzyme or motor, following U. SEIFERT, 2005, as depicted in Figure 2.1. The enzyme is assumed to have three equivalent conformational states, and it progresses from one to the other at a rate k_+ in the forward (i.e., clockwise), and with a rate k_- in the backward direction. We assume $k_+ > k_-$. The stationary distribution of the system clearly is given by $p^s = 1/3$: the system spends equal times in each of the states. For this system we obtain R as

$$R = n \ln(k_+/k_-) \qquad (2.63)$$

where $n \equiv n_+ - n_-$ is the effective number of steps in the forward direction. Thus

$$\frac{P[-n]}{P[n]} = e^{-n \ln(k_+/k_-)} = \left(\frac{k_-}{k_+}\right)^n . \qquad (2.64)$$

The exact $P[n]$ can be computed from the master equation, since this system is an *asymmetric random walk* with the master equation

$$\dot p_n = k_+ p_{n+1} - k_- p_{n-1} - (k_+ + k_-)p_n \qquad (2.65)$$

for which

$$p_n \equiv P[n] = I_{|n|}(2\sqrt{k_+k_-}t) \left(\frac{k_+}{k_-}\right)^{n/2} e^{-(k_+ + k_-)t} , \qquad (2.66)$$

where $I_n(x)$ is the modified Bessel function of order[8] n. One sees that the factor $(k_-/k_+)^n$ arises from the ratio of backwards and forward processes with the probability ratio $P[-n]/P[n]$.

---

[8] The modified Bessel function of order n is given by the expression

$$I_\alpha(x) = i^{-\alpha} J_\alpha(x)$$

with

$$J_\alpha(x) = \sum_{m=0}^\infty \frac{(-1)^m}{m!\Gamma(m+\alpha+1)}\left(\frac{x}{2}\right)^{2m+\alpha} .$$

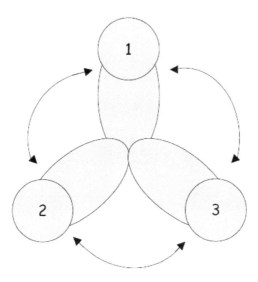

FIGURE I.2.1: An enzyme or molecular motor switching between three configurational states. Transition rates in forward and backward direction are given by k_i^+ and k_i^- for $i = 1, 2, 3$, respectively (after U. SEIFERT, 2005).

Jarzynszki theorem. The fluctuation theorem above can further be elucidated by pointing out its relation of eq.(2.62) with a theorem due to C. JARZYNSKI, 1997, as suggested by U. SEIFERT, 2004.

Consider the trajectory of a particle $x(t)$ in a potential $V(x, \lambda)$, which again depends on the parameter λ. The stationary distribution of this process for fixed λ is given by

$$p^s(x, \lambda) = Z(\lambda)^{-1} \exp[-\beta V(x, \lambda)] \tag{2.67}$$

with the normalization given by the 'partition function'

$$Z(\lambda) \equiv \int_{-\infty}^{\infty} dx \exp[-V(x, \lambda)]. \tag{2.68}$$

Substituting these correspondences into the formula (2.59) one has with $n(t) \sim x(t)$

$$\epsilon_n(\lambda) \sim -\ln p^s(x, \lambda) = \beta V(x, \lambda) + \ln Z(\lambda). \tag{2.69}$$

Inserted into eq.(2.62) this leads to the result

$$\langle \exp\left(-\beta \int_0^t d\tau V'(x(\tau), \dot{\lambda}(\tau))\right) \frac{Z(\lambda(0))}{Z(\lambda(t))}\rangle = 1. \tag{2.70}$$

Invoking the expression for the free energy

$$F(\lambda) = \beta^{-1} \ln Z(\lambda), \tag{2.71}$$

and introducing the notions of the *applied work*

$$W \equiv \int_0^t d\tau V'[x(\tau), \lambda(\tau)]\dot{\lambda}(\tau) \tag{2.72}$$

and the *dissipated work*

$$W_{diss} \equiv W - [F(\lambda(t)) - F(\lambda(0))] \tag{2.73}$$

we finally arrive at the expression

$$\langle e^{-\beta W_d}\rangle = 1 \tag{2.74}$$

or, equivalently, at

$$\langle e^{-\beta W}\rangle = e^{-\beta \Delta F} \tag{2.75}$$

where ΔF can be read off from eq.(2.73). This is Jarzynski's original result (1997).

The Jarzynski equation (2.75) is remarkable since it relates the difference between an *equilibrium free-energy difference* to the average over a *nonequilibrium quantity*. This merits a deeper discussion.

Validation of Jarzynski's equality. Jarzynski's result can be rewritten as an equation for the free energy difference in terms of a nonequilibrium expression

$$\Delta F = \beta^{-1} \ln\langle e^{-\beta W}\rangle. \tag{2.76}$$

To interpret this result further it is useful to rewrite it further in terms of the *cumulant expansion* (see Chapter 1). This operation leads to the expression

$$\Delta F = \sum_{n=1}^{\infty} \frac{1}{n!}\kappa_n(-\beta)^{n-1} \tag{2.77}$$

where the first four cumulants are given by

$$\kappa_1 = \langle W\rangle \ , \quad \kappa_2 = \langle (W - \langle W\rangle)^2\rangle = \sigma_W^2 \tag{2.78}$$

and

$$\kappa_3 = \langle (W - \langle W \rangle)^3 \rangle \; , \quad \kappa_4 = \langle (W - \langle W \rangle)^4 \rangle - 3\sigma_W^4 \; . \tag{2.79}$$

These expressions are instructive since we can infer from them different levels of approximation which are testable against experimental measurements or simulation results.

To begin, we keep only the first term of the cumulant expansion. We then estimate the work done by the system as the free energy difference. This is correct only when the work is done *reversibly*, which means that there is no mechanism of energy dissipation in the system.

The second level approximation amounts to consider

$$\Delta F = \langle W \rangle - \frac{1}{2}\beta \sigma_W^2 \; , \tag{2.80}$$

i.e., the fluctuations around the stationary state. Eq.(2.80) becomes exact in a regime near an equilibrium state, since then the work distribution is Gaussian, and all higher cumulants vanish identically. The result is also an example application of the *fluctuation-dissipation theorem*, since

$$\overline{W}_{diss} = \frac{1}{2}\beta \sigma_W^2 \; , \tag{2.81}$$

relates the dissipated work to the Gaussian fluctuations.

If we want to take the full Jarzynski result serious, we have to estimate the free energy difference by the following expression

$$\Delta F = -\frac{1}{\beta} \ln \left[\frac{1}{N} \sum_{i=1}^{N} e^{-\beta W_i} \right] \; , \tag{2.82}$$

where N is the number of trajectories for which the work W has been determined.

What trajectories are best to measure in an experiment or a simulation? This can be understood from reconsidering the Jarzynski-result in the form involving the dissipated work. Since we are near an equilibrium state, W is Gaussian-distributed, hence the dissipated work follows a Gaussian distribution with mean $\overline{W}_{diss} = \frac{1}{2}\beta \sigma_W^2$ and variance σ_W^2. As $\overline{W}_{diss}$ increases - one moves away from the near-equilibrium regime - the distribution will broaden. The expression eq.(2.82) obviously heavily weighs those trajectories whose dissipated work is *negative*. The probability of finding such a trajectory is given by

$$P(W_{diss} < 0) = \int_{-\infty}^{0} dW_{diss} P(W_{diss}) = \frac{1}{2}(1 - erf(\sqrt{\overline{W}_{diss}}/2)) \tag{2.83}$$

where erf is the error function.[9] Since eq.(2.83) is a sharply decreasing function of its argument, it is established that the efficiency of sampling falls rapidly with the increase of dissipated work.

Jarzynski's equality has been tested in experiments on pulling RNA secondary structure (for details on RNA secondary structure, see Part II of the book). In such experiments - using a similar experimental setup we described in Chapter 1 in the context of the Worm-Like Chain model - a single RNA molecule with known secondary structure is fixed at its extremeties between two beads which can be moved reversibly, at different speeds (J. LIPHARDT et al., 2002). These experiments have indeed allowed to demonstrate that the estimator given by eq.(2.82) converges towards ΔF if sufficiently many trajectories N are taken into account for which the dissipated work is high. The quantity can also be validated in simulations (S. PARK et al., 2003). Again, a sufficient sampling range is needed. More recently, a detailed study of RNA folding free energies has been performed by D. COLLIN et al., 2005, verified a generalization of the Jarzynski theorem, the *Crooks fluctuation theorem*, in regimes near and far from equilibrium. In the context of RNA folding, this theorem states that the probability the ratio of the probability distributions of the work for unfolding and folding under conditions of time-reversal symmetry fulfills

$$\frac{P_{unfold}(W)}{P_{fold}(W)} = \exp \beta(W - \Delta F) . \qquad (2.84)$$

2.4   The Fokker-Planck and Langevin equations

In this section, we discuss two further classic analytical approaches commonly used to describe the stochastic dynamics of particle systems.

The Fokker-Planck equation. In this subsection we will introduce a continuous approximation to the master equation. First, we recall the last general expression we had obtained for it, eq.(2.24)

$$\partial_t P(y,t) = \int dy' \left[W(y|y')P(y',t) - W(y'|y)P(y,t) \right] . \qquad (2.85)$$

---

[9]The error function is defined by

$$erf(x) \equiv \frac{2}{\sqrt{\pi}} \int_0^x dt e^{-t^2} .$$

We now rewrite the transition probabilities W as

$$W(y|y') = W(y';r) , \quad r = y - y' \qquad (2.86)$$

and obtain

$$\partial_t P(y,t) = \int dr W(y-r;r)P(y-r,t) - P(y,t) \int dr W(y;-r). \qquad (2.87)$$

This is just a rewrite, but now we want to play with the difference variable $r = y - y'$ which will allow us to formulate a continuum approximation.

In order to formulate this continuum version of the master equation we assume for the dependence of the transition probabilities on r that they fulfill the following

$$W(y';r) \approx 0, \quad |r| > \delta \qquad (2.88)$$

$$\qquad (2.89)$$

$$W(y' + \Delta y;r) \approx W(y';r), \quad |\Delta y| < \delta \qquad (2.90)$$

which means that W is a slowly varying function of y', but sharply peaked in its dependence on r. If, additionally, P varies also slowly with y, we can expand in a Taylor series to obtain

$$\partial_t P(y,t) = \int dr W(y;r)P(y;t) - \int dr\, r\, \partial_y[W(y;r)P(y,t)]$$

$$\qquad (2.91)$$

$$+ \frac{1}{2} \int dr\, r^2 \partial_y^2[W(y;r)P(y,t)] - P(y,t) \int dr W(y;-r)$$

where the first and the last term on the rhs cancel each other out. The remaining integrals over r can be absorbed in the definition of the *jump moments*

$$a_\alpha = \int_{-\infty}^{\infty} dr r^\alpha W(y;r) \qquad (2.92)$$

and we finally obtain

$$\partial_t P(y,t) = -\partial_y[a_1(y)P(y,t)] + \frac{1}{2}\partial_y^2[a_2(y)P(y,t)]. \qquad (2.93)$$

This is the *Fokker-Planck equation*. It contains a *drift term* $\sim \partial_y P$ and a *diffusion term*, $\sim \partial_y^2 P$. Remember that we had introduced this terminology before in the context of the general formulation of the dissipation-fluctuation theorem. The Fokker-Planck equation makes this relationship again explicit: if we suppose a stationary distribution, $\partial_t P = 0$, drift and diffusion term have to balance each other.

Task. Rederive the dissipation-fluctuation theorem of Section 2.3 for the Fokker-Planck equation.

Example for the Fokker-Planck equation: Brownian motion. Consider a particle suspended in a liquid. If we trace its motion under the influence of random molecular collisions of the liquid, we find a continuous path, as shown in Figure 2.2. Measuring the distances the particle travels, and averaging over several realizations we obtain for the jump moments the expressions

$$a_1 \equiv \frac{\langle \Delta x \rangle_X}{\Delta t} = 0, \quad a_2 \equiv \frac{\langle (\Delta x)^2 \rangle_X}{\Delta t} = const. \tag{2.94}$$

Thus, the Brownian particle obeys a diffusion equation

$$\partial_t P(x,t) = \frac{a_2}{2} \partial_x^2 P(x,t) \tag{2.95}$$

and we can identify $a_2/2 \equiv D$, i.e. the diffusion coefficient is given by

$$D = \frac{\langle (\Delta x)^2 \rangle}{\Delta t}. \tag{2.96}$$

We now want take a second, different look at the Brownian particle.[10] Let us now consider the velocity instead of the position of a suspended particle as the dynamic variable. We first ignore the random collisions of the particle of the solution molecules and assume that the velocity of the particle relaxes according to

$$\dot{v} = -\gamma v, \tag{2.97}$$

i.e., the particle velocity will go to zero for $t \to \infty$: the differential equation is the same as that for the mean number of proteins in the protein degradation problem we discussed with help of the master equation.

Hence, there is a now a drift term in the corresponding Fokker-Planck equation, and it is given by the jump moment

---

[10]What we consider here is actually called the *Rayleigh particle* which is equivalent to the Brownian particle; the difference between the two is the fine graining of the time scale. In the discussion of the Rayleigh particle one assumes that $t_{coll} \ll \Delta t \ll t_{relax}$, where t_{coll} is a molecular collision time, and t_{relax} the time scale on which the particle velocity relaxes.

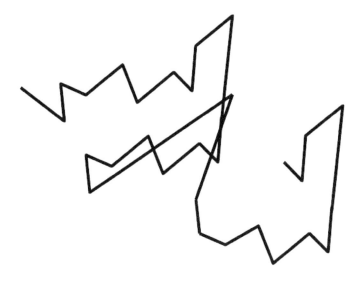

FIGURE I.2.2: Particle trajectory in solution: a Brownian path.

$$a_1(v) = \frac{\langle \Delta v \rangle_V}{\Delta t} = -\gamma v. \tag{2.98}$$

The second jump moment fulfills $a_2(v) > 0$ even for $v = 0$. The Fokker-Planck equation reads as

$$\partial_t P(v,t) = \gamma \partial_v [v P(v,t)] + \frac{a_2}{2} \partial_v^2 P(v,t). \tag{2.99}$$

The stationary distribution for this process is actually known from equilibrium statistical mechanics: it is the Maxwell-Boltzmann velocity distribution in which m is the particle mass and $\beta = 1/k_B T$,

$$P(v) = \left(\frac{\beta m}{2\pi}\right)^{1/2} \exp\left(-\frac{\beta m v^2}{2}\right). \tag{2.100}$$

With the help of this expression, we can identify the coefficient of the diffusion term as $a_2/2 = \gamma/(m\beta)$, and the Fokker-Planck equation is fully given by

$$\partial_t P(v,t) = \gamma \partial_v \left[v P(v,t) + \frac{1}{\beta m} \partial_v P(v,t) \right]. \tag{2.101}$$

Exercise. Compute $\langle v(t) \rangle$ and $\langle v^2(t) \rangle$ from eq.(2.101) for the given initial velocity $v(0) = v_0$.

Exercise. Compute the dissipation-fluctuation relation for the Brownian motion from the Fokker-Planck equation for $P(v,t)$.

The Langevin equation. We now complete the discussion of the Brownian motion of the suspended particle in the velocity description by the classic approach originally suggested by P. LANGEVIN.

In order to do this we consider the velocity process under the inclusion of a noise source $\eta(t)$ which we add to the rhs of equation (2.97),

$$\dot{v} = -\gamma v + \eta(t). \tag{2.102}$$

The noise source η models the random action of the solution molecules on the suspended particle (we now put $m = 1$). The solution particles 'kick' the suspended particle and transfer part of their thermal energy. For the first two moments of the noise distribution we assume

$$\langle \eta(t) \rangle = 0, \quad \langle \eta(t)\eta(t') \rangle = \Gamma\delta(t - t'), \tag{2.103}$$

where Γ is a constant. The motivating idea behind this specific choice is that the random collisions are instantaneous and uncorrelated.

Exercise. How would the rhs of the second moment change if one were to introduce a small but finite collision time τ_c?

If we assume an initial velocity $v(0) = v_0$ of the particle, as we did before, we can compute the velocity $v(t)$ as

$$v(t) = v_0 e^{-\gamma t} + e^{-\gamma t} \int_0^t dt e^{\gamma t'} \eta(t') \tag{2.104}$$

and, using the moments of η, we can obtain those of v as

$$\langle v \rangle = v_0 e^{-\gamma t} \tag{2.105}$$

$$\langle v^2(t) \rangle = v_0^2 e^{-2\gamma t} + \frac{\Gamma}{2\gamma}(1 - e^{-2\gamma t}). \tag{2.106}$$

We are now left to determine the coefficient Γ. For $t \to \infty$, we know from eq.(2.100) - for $m = 1$ - that

$$\langle v^2(t \to \infty) \rangle = k_B T = \frac{\Gamma}{2\gamma}. \tag{2.107}$$

Again we have found an example of the *dissipation-fluctuation theorem*. A further, very prominent example is the *Einstein relation*

$$D = \frac{k_B T}{\gamma},$$
(2.108)

see the *Exercise* on the Fokker-Planck equation in velocity space.

Fokker-Planck vs. Langevin. The Fokker-Planck equation determines the full stochastic process of the Brownian particle; by contrast, the Langevin equation does, by construction, not go beyond second moments. Therefore, if we additionally assume that the noise η is Gaussian, all odd moments will vanish, and the even moments will factorize (*Exercise!*). This leads to what is commonly called a *Gaussian white noise*, i.e., a noise spectrum containing all frequencies, which usually serves to model a rapidly fluctuating force.

This construction is all fine for the simple case of a Brownian particle, and lends itself to the many applications of the Langevin equation approach, in which a Gaussian white noise is added to the known deterministic equations of the system. The constant Γ is then usually adjusted such that the stationary solution correctly matches with the fluctuations around the stationary state.

But it is worth keeping in mind that this approach is a strong simplification, as has been advocated by N. VAN KAMPEN. We will briefly go through the main points here; for a more detailed discussion the reader is referred to van Kampen's book. Furthermore, we will return to this discussion at the end of Part III when we will discuss the role of fluctuations in biological systems.

What has to be kept in mind if one wants to treat stochastic fluctuations with the Langevin approach? By this we refer to the procedure we used for the description of Brownian motion, namely to first write down a deterministic equation for the macroscopic (average) dynamics, and to then add on the fluctuations.

i) Suppose your system is described by a deterministic equation of the type $\dot{u} = A(u) = \sin u$ to which we add the noise as defined above,

$$\partial_t u = \sin u + \eta(t).$$
(2.109)

If we average this equation, we find

$$\partial_t \langle u \rangle + \langle \sin u \rangle = 0$$
(2.110)

since, as before, we have $\langle \eta(t) \rangle = 0$. This result, however, means that the average does not obey the macroscopic equation, i.e., $\partial_t \langle u \rangle \neq \sin \langle u \rangle$, since

$$\langle \sin u \rangle = \sin \langle u \rangle - \frac{1}{2} \langle (u - \langle u \rangle)^2 \rangle \cos \langle u \rangle + \dots, \tag{2.111}$$

which is an equation involving all higher moments.

The message of this calculation obviously is: if we start from a deterministic equation for the average which is nonlinear, the simple addition of fluctuating source will in general be too naive.

ii) For an arbitrary nonlinearity $A(u)$ the Langevin equation eq.(2.109) is equivalent to the Fokker-Planck equation

$$\partial_t P(u,t) = -\partial_u (A(u)P(u,t)) + \frac{\Gamma}{2} \partial_u^2 P(u,t). \tag{2.112}$$

If we allow equations of the type

$$\partial_t u = A(u) + C(u)\eta(t) \tag{2.113}$$

in which a u-dependent function multiplies the noise, we run into an inter-pretation problem. For each jump in the solution u of the equation due to the noise, the value of u and hence of $C(u)$ is undetermined. We thus have to specify a rule how to interpret the product $C(u)\eta(t)$, and this leads to a dependence of the resulting Fokker-Planck equation on that rule. Some pos-sibilities are to take the value of $C(u)$ before the jump, after the jump or the mean. The different options lead to different Fokker-Planck equations and hence to different results.[11] The option to take the mean value is named after R. L. STRATONOVICH, the version to take the value before the jump is named after K. ITÔ.

We illustrate this phenomenon for the example of protein degradation for which we had the master equation

$$\dot{p}_n = \gamma(n+1)p_{n+1}(t) - \gamma n p_n(t). \tag{2.114}$$

Within the Langevin approach we can assume

$$\dot{n} = -\gamma n + \sqrt{n}\eta(t) \tag{2.115}$$

since we expect the fluctuations in the decay process to be proportional to the square root of the number of proteins. Thus, we have a case in which we

---

[11]The two stochastic equations differ by what is called a *spurious drift term*, and the equations can be transformed into each other by corresponding transformation rules; see the detailed discussion in van Kampen's book.

need to determine the multiplication rule for the noise.

In the Itô-case, we take the value before the jump, hence

$$n(t + \Delta t) - n(t) = -\gamma n(t)\Delta t + \sqrt{n} \int_t^{t+\Delta t} dt' \eta(t') \, . \tag{2.116}$$

It can be proved that this choice leads to the Fokker-Planck equation

$$\partial_t P(n, t) = \partial_n (\gamma + \frac{\Gamma}{2}\partial_n)[nP(n, t)] \, . \tag{2.117}$$

The corresponding line of reasoning in the Stratonovich case yields

$$\partial_t P(n, t) = \partial_n \left[\left(\gamma n - \frac{\Gamma}{2} \right) P + \frac{\Gamma}{2}\partial_n[nP(n, t)] \right] \, . \tag{2.118}$$

If we calculate the equation for the average $\langle \dot{n} \rangle$ from the Itô-version of the Fokker-Planck equation, we obtain the same result as eq.(2.28) which we obtained directly from the master equation (*Exercise*). The Stratonovich equation, by contrast, does not yield this result. The reason is that in the process of protein degradation, the probability to go from n to $n-1$ proteins is indeed proportional to the number n of available proteins before the transition, in line with the Itô assumption of the construction of the stochastic process in the first place.

The ambiguity arising from the different possible choices is referred to as the *Itô-Stratonovich dilemma*. A better understanding of what is an adequate procedure in formulating stochastic equations can be obtained from a distinction between *intrinsic* and *extrinsic* noise sources. Extrinsic noise is an 'add-on' to a fundamentally deterministic dynamics, as it is the case in many engineering applications. Here, one models, e.g., an electrical circuit on the level of its macroscopic constituents like resistors and capacitors and not on the level of electrons. In such a case, noise is never really white but has a finite correlation time. In this case, there is no ambiguity and the Stratonovich prescription applies.

Intrinsic noise, by contrast, is due to the fact that the system itself is composed of discrete particles which interact stochastically: this noise can *never* be switched off, and hence $A(u)$ is not determined by the evolution equations in a system isolated from noise. Hence, the Langevin approach does in general not work for internal noise. Instead, one has to go back to the master equation approach, from which in certain cases macroscopic equations can be derived.

This derivation can be done as follows in a systematic way. The starting point is the probability density $P(y,t)$ for a stochastic process Y_X. We define

$$y(t) \equiv \langle Y \rangle_t = \int dy \, y \, P(y,t) \tag{2.119}$$

where we assume that the density P is a sharply peaked function of its argument y; it is assumed to be initially of the form $P(y,0) = \delta(y - y_0)$.

As t increases, the probability density P will evolve along y, and this determines the behaviour of the average $y(t)$. One finds

$$\dot{y}(t) = \int dy \, y \, \partial_t P(y,t)$$

$$= \int dy \, dy' y \, [W(y|y')P(y',t) - W(y'|y)P(y|t)] \tag{2.120}$$

$$= \int dy \, dy' (y' - y)W(y'|y)P(y,t) \, . \tag{2.121}$$

We now recall the *jump moments* (eq.(2.92))

$$a_\alpha(y) = \int dy' (y' - y)^\alpha W(y'|y) \tag{2.122}$$

for $\alpha = 0, 1, 2, ..,$ to obtain

$$\dot{y}(t) \equiv \frac{d}{dt}\langle Y \rangle_t = \int dy \, a_1(y)P(y,t) = \langle a_1(Y) \rangle_t \tag{2.123}$$

If we expand $a_1(y)$ around the average $\langle Y \rangle_t$ in a Taylor series

$$\langle a_1(Y) \rangle_t = a_1(\langle Y \rangle_t) + \frac{1}{2}\langle (Y - \langle Y \rangle_t)^2 \rangle_t a_1''(\langle Y \rangle_t) + ... \tag{2.124}$$

one finds to a first approximation the equation

$$\dot{y}(t) = a_1(y(t)) = A(y) \tag{2.125}$$

which is, despite P fulfilling a linear equation, in general a *nonlinear* equation of $y(t)$, even when the higher order terms are dropped.

Task. Compute the corrections to the macroscopic equations. For this, consider

$$\frac{d}{dt}\langle Y^2(t) \rangle_t \tag{2.126}$$

and the variance $\sigma^2(t) = \langle Y^2(t) \rangle - y^2(t)$. For jump moments a_1 and a_2 which are linear in y show that the macroscopic equation can be written as

$$\dot{y}(t) = a_1(y(t)) + \frac{1}{2}\sigma^2(t)a_1''(y(t)) . \tag{2.127}$$

This concludes our discussion of the general methods of stochastic dynamics, i.e., master equations, Fokker-Planck and Langevin equations. We now move on to a class of systems to which these approaches will be applied later, to biochemical reaction systems, and give the basic elements of the theory of chemical kinetics. We discuss two approaches: first, the deterministic one, based on rate equations. Subsequently, we return to the stochastic aspect by a description of the Gillespie algorithm which allows to simulate the master equation for biochemical reactions.

Chemical kinetics. In many applications of stochastic processes in biology, we have to deal with biochemical reactions - we will see such examples in Part III of the book.

Consider a system of m chemical substances, and n chemical reactions. We represent the *chemical reactions* they undergo by n *collision diagrams*, given by

$$\nu_{i1}A_1 + ... + \nu_{im}A_m \xrightarrow{k_i} \mu_{i1}A_1 + ... + \mu_{im}A_m \tag{2.128}$$

for $i = 1, .., n$. A_i can represent either the number of a molecule of substance i, or its concentration (obtained by division through volume, $n_i = A_i/\Omega$). The k_i are *rate constants*; ν_{im}, μ_{im} are stochiometric coefficients, which are integer numbers. The sums

$$r_i = \sum_{j=1}^{m} \nu_{ij} \tag{2.129}$$

define the *order of the transitions*. A transition

$$A_1 + A_2 \xrightarrow{k} A_3 \tag{2.130}$$

thus represents the binary collision of A_1 and A_2, giving rise to A_3. Higher order collisions have a low probability to occur, and are generally neglected in chemical kinetics.

In order to pass from a *collision diagram* to equations describing the time evolution, several assumptions have to be made. We begin with the following:

- the system (typically, a solution) can be assumed spatially homogeneous;

- the density of the molecules is low;[12]

- all reactions occur at a constant volume and temperature;

- the collision frequency of the molecules depends on local concentration.

We can then write down the equation for the chemical kinetics of the molecules in the form

$$\frac{dA_j}{dt} = \sum_{i=1}^{n} k_i (\mu_{ij} - \nu_{ij}) A_1^{\nu_{i1}} \cdots A_m^{\nu_{im}} \tag{2.131}$$

for $j = 1, ..., m$. This expression is the *rate equation*.

The equations expressed in (2.131) are not always independent. This can be made more explicit by considering them in matrix form

$$\frac{d}{dt} \mathbf{A} = \mathbf{M} \cdot \mathbf{K} \tag{2.132}$$

where A_j is a vector with components A_j, $\mathbf{M}$ is a matrix with row vectors $\mathbf{V}^T \equiv (\mu_{1j} - \nu_{1_j}, ..., \mu_{nj} - \nu_{nj})$ for $j = 1, ..., m$, and $\mathbf{K}^T \equiv (k_1 A_1^{\nu_{i1}} \cdots A_m^{\nu_{1m}}, ..., k_n A_n^{\nu_{n1}} \cdots A_m^{\nu_{nm}})$. If the matrix $\mathbf{M}$ has a rank r with $r \leq \min\{n, m\}$ the equations are linearly dependent; there are thus $m - r$ *conservation laws* which can be written in the form

$$\sum_{j=1}^{m} \alpha_{jk} A_j(t) = \sum_{j=1}^{m} \alpha_{jk} A_j(0) \tag{2.133}$$

for $k = 1, ..., m - r$.

Task. Consider eq.(2.131) for the case of one collision diagram, with reaction and back-reaction rate constants k_+ and k_-.

a) Write down the *chemical master equation* using the step operator $\mathcal{E}$.

b) For an ideal mixture of molecules, the grand canonical distribution is given by

$$P(\{n_i\}) = \prod_j \frac{(\Omega z_j)^{n_j}}{n_j!} \exp(-\Omega z_j) \tag{2.134}$$

with $n_j = 0, 1, ...$ Here

$$z_j = \left(\frac{2\pi m}{\beta} \right)^{3/2} \sum_\nu \exp(-\beta \epsilon_\nu) \tag{2.135}$$

---

[12]But not too low; if the molecules appear in too few numbers, the continuum approach breaks down.

is the partition function of one molecule j in a unit volume. The energies ϵ_ν contain all internal molecular degrees of freedom, be they rotational, vibrational or electronic. Check that $P(\{n_i\})$ is a stationary solution of the master equation obtained under a), if

$$\frac{k_+}{k_-} = \prod_j z_j^{\mu_j - \nu_j} \tag{2.136}$$

applies, i.e., the *law of mass action* is fulfilled.

We will turn to the discussion of specific examples of kinetic equations in Part III of the book.

The Gillespie algorithm. Since master equations are usually easy to write down but hard to solve explicitly, D. T. GILLESPIE, 1977, proposed a simple prescription how to determine the time evolution corresponding to the master equation from a stochastic algorithm.

Starting point of the algorithm is the expression for the transition probability of the Chapman-Kolmogorov equation, eq.(2.21), in the case of small time interval τ. The algorithm can be formulated in terms of the probability P for a given reaction $i = 1, ..., n$ to occur in an infinitesimal time-interval $d\tau$, which can be written as

$$P(\tau, i) = P_0(\tau) \cdot a_i \cdot d\tau \tag{2.137}$$

where $P_0(\tau)$ is the probability that *no* collision will have occurred in the interval $(t, t + \tau)$, while $a_i \equiv h_i c_i$ is a *stochastic reaction rate* c_i, multiplied by a combinatorial factor determined by the type of reaction that occurs, and by counting the number of distinct reaction partners that are available for a reaction. Thus, $a_i \cdot d\tau$ is the probability that a collision will occur in the interval $(t + \tau, t + \tau + d\tau)$. Defining

$$a_0 \equiv \sum_{i=1}^{M} a_i , \tag{2.138}$$

$P_0(\tau)$ obeys the equation

$$P_0(\tau + d\tau) = P_0(\tau) \cdot [1 - a_0 \cdot d\tau] \tag{2.139}$$

which describes the (obvious) fact that the probability that *no* reaction occurred in time $\tau + d\tau$ equals the product of the probabilities that no transition occured in the interval τ, and within $d\tau$.

The solution of this equation is given by $P_0(\tau) = \exp(-a_0\tau)$, such that

$$P(\tau; i) = a_i \exp(-a_0\tau), \quad 0 \leq \tau < \infty, i = 1, .., n, \qquad (2.140)$$

or 0 otherwise.

This result can be used to compute the time evolution of the system by the reconstruction of $P(\tau; i)$ from a draw of two random numbers r_1 and r_2 from the unit interval uniform distibution according to the prescription

$$\tau = \frac{1}{a_0} \ln\left(\frac{1}{r_1}\right), \quad i = r_2 \qquad (2.141)$$

where i is selected such that the condition

$$\sum_{j=1}^{i-1} a_j < r_2 a_0 \leq \sum_{j=1}^{i} a_j \qquad (2.142)$$

is fulfilled.

These steps can finally be cast into the following algorithm:

- define the stochastic rates c_i for $i = 1, ..., n$, and the initial conditions on the N molecules;

- calculate $a_1 = h_1 c_1, ..., a_n = h_n c_n$, and $a_0 = \sum_{j=1}^{n} a_j$;

- generate r_1, r_2 and calculate τ and i;

- increase t by τ, adjust the population levels of the selected reaction i; loop.

2.5   Sequence alignment: a nonequilibrium phase transition

Now we are fully done with the basic methods of stochastic processes. In the last section of Part I of this book we will now discuss an example which draws from all aspects that were discussed in the two Chapters of this Part: we will discuss a phase transition, but not an equilibrium one, we will find critical exponents and scaling, and we will both use stochastic equations in their discrete and continuous versions. The topic is *sequence alignment*, a fundamental method in computational biology developed to find out the similarities between two sequences, be they made up DNA bases, or the amino acid sequences of proteins.

In this section, we will see that the alignment of two sequences can be understood as a phase transition in a nonequilibrium system. This may, at first sight, seem astonishing: what does a pattern matching problem have to do at all with a phase transition? There are several aspects that have to be addressed in order to answer this question, and we will do so as we go along, and follow the exposition by R. BUNDSCHUH, 2002.

We begin with a technical definition of the notion of an alignment.

Gapless alignment. The simplest procedure of sequence alignment is called *gapless alignment*. It looks for similarities between two sequences $\mathbf{a} = (a_1, a_2, ..., a_M)$ and $\mathbf{b} = (b_1, b_2, ..., b_N)$ where $M \sim N$. The letters a_i, b_i are taken from an alphabet of size c; for our purposes here we take $c = 4$, i.e., the four-letter alphabet of the DNA bases.

A local gapless alignment A consists of two substrings of equal length ℓ of the sequences $\mathbf{a}, \mathbf{b}$. To each such alignment can be assigned a *score*

$$S[A] = S(i, j, \ell) = \sum_{k=0}^{\ell-1} s_{a_{i-k}, b_{j-k}} \tag{2.143}$$

where the *scoring matrix* is given by in the simplest case

$$s_{a,b} = \begin{cases} 1 & a = b \\ -\mu & a \neq b \end{cases}. \tag{2.144}$$

What is to be computed are the values of i, j and ℓ which lead to the highest total score

$$\Sigma \equiv \max_A S[A] \tag{2.145}$$

for the given scoring matrix $s_{a,b}$.

This optimization problem can be reformulated by introducing the auxiliary quantity $S_{i,j}$ which is the optimal score of the subsequences ending at (i, j) optimized over ℓ. This quantity can be computed with $O(N^2)$ steps instead of $O(N^3)$ with the prescription

$$S_{i,j} = \max\{S_{i-1,j-1} + s_{a_i,b_j}, 0\} \tag{2.146}$$

for the initial condition $S_{0,k} = S_{k,0} = 0$. This recursion expresses the fact that for a given pair (i, j) the optimal $\ell = 0$ or $\ell > 0$. If $\ell = 0$, the resulting score is zero either; if $\ell = 1$ at least, the corresponding pair (a_i, b_j) will belong to the optimal alignment (which may be longer), whatever had been chosen as optimal up to $(i - 1, j - 1)$. Eq.(2.146) describes a random walk with increment $s_{a,b}$. It is cut off when it falls below zero. The global score Σ is then given by

$$\Sigma = \max_{1 \le i \le M, 1 \le j \le N} S_{i,j} \,. \tag{2.147}$$

Significance of the alignment. Suppose we have found an alignment by performing this computation. How significant is it? In order to answer this question, we have to discuss the alignment of purely random sequences, and to determine the distribution of scores in this case. This can be done rigorously, leading to the result

$$P[\Sigma < S] = \exp(-\kappa e^{-\lambda S}) \,, \tag{2.148}$$

which is a *Gumbel* or *extreme value distribution*. It is characterized by two parameters λ and κ where λ characterizes the tail of the distribution. For the case of gapless alignment we discuss here, both parameters can be calculated from the knowledge of the scoring matrix $s_{a,b}$.

We want to illustrate how the result (2.148) comes about in a heuristic fashion. For this we go back to equation (2.146) and set $i = j$, which is permissible for random sequences. Thus we have

$$S_{i,i} \equiv S(i) = \max\{S(i - 1) + s(i), 0\} \,. \tag{2.149}$$

In this equation, $s(i) = s_{a,b}$ plays the role of an uncorrelated noise given by the distribution

$$P[s(i) > s] = \sum_{(a,b|s_{a,b} > s)} p_a p_b \,. \tag{2.150}$$

Eq.(2.149) is essentially a discrete version of a Langevin equation. The dynamics it generates can be found to be in two distinct phases, called the *linear* and the *logarithmic phase*. The quantity which distinguishes the two is the local similarity score

$$\langle s \rangle = \sum_{(a,b)} p_a p_b \,. \tag{2.151}$$

In the linear phase, the dynamics is a random walk with an average upward drift $\langle s \rangle$, and the maximal score is $\Sigma \approx N \langle s \rangle$ for a sequence of length N. This phase constitutes a phase of *global alignment*, and hence does not permit to identify similarities in subsequences. The distribution of the Σ in this phase is Gaussian, not Gumbel. If, however, the average drift $\langle s \rangle$ is negative, the ensuing dynamics will lead to a score of the form shown in Figure 2.3. The resulting score landscape can be considered as consisting of 'islands' in an

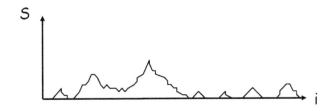

FIGURE I.2.3: Total score of an alignment as a function of sequence position (after R. BUNDSCHUH, 2002)

'ocean'. For the case of random sequences, the islands are statistically independent. If σ_k is the maximal score of island k, the σ_k are thus independent random variables.

Island distribution. In order to calculate the island distribution explicitly, we have to look at (R. BUNDSCHUH, 2002)

$$p(\sigma) = \langle \delta(\sigma - \sum_{i=1}^{L} s(i)) \rangle. \tag{2.152}$$

Here L is the length of a 'large' island, measured from its beginning to the peak at height σ. Using the Fourier representation of the δ-function and the statistical independence of the $s(i)$ we obtain

$$p(\sigma) = \frac{1}{2\pi} \int dk e^{-ik\sigma} \langle e^{iks} \rangle^L \tag{2.153}$$

Assuming that the peak score of each island is proportional to its length - islands are thus on average little triangles with rising slope α - we approximate

$$p(\sigma) \approx \frac{1}{2\pi} \int dk e^{-ik\alpha L} \langle e^{iks} \rangle^L \tag{2.154}$$

and then evaluate the integral in a saddle-point approximation. This leads to (*Task!*)

$$p(\sigma) \sim \exp(-\lambda\sigma) \tag{2.155}$$

with

$$\lambda = ik_s - \ln[\langle e^{ik_s s} \rangle]/\alpha, \tag{2.156}$$

where the saddle-point value of k, which we call k_s, is determined by

$$\frac{\langle s e^{ik_s s} \rangle}{\langle e^{ik_s s} \rangle} = \alpha. \tag{2.157}$$

Note that k_s still depends on α. The correct value α is found by minimizing eq.(2.157) with respect to α and using eq.(2.157), which yields

$$\langle e^{ik_s s} \rangle = 1 \tag{2.158}$$

i.e., explicitly

$$\langle e^{\lambda s} \rangle = \sum_{a,b} p_a p_b e^{\lambda s_{a,b}} = 1 \,. \tag{2.159}$$

The typical slope of an island is given by

$$\alpha = \langle s e^{\lambda s} \rangle \,. \tag{2.160}$$

Thus, we conclude that the islands follow the exponential distribution

$$P[\sigma_k > \sigma] \approx C e^{-\lambda \sigma} \,. \tag{2.161}$$

Since the global optimal score Σ is given in terms of the islands σ_k by

$$\Sigma = \max_{k}\{\sigma_k\} \tag{2.162}$$

the distribution of the Σ can be computed from the distribution of the σ_k. For a large number $K \sim N$ of island peaks one finds

$$P[\Sigma < S] = P[\max\{\sigma_1, .., \sigma_K\} < S] = [1 - C e^{-\lambda S}]^K \approx exp(-\kappa e^{-\lambda S}), \tag{2.163}$$

where $\kappa = CK$.

Gapped alignment. We now turn to the case of alignment with gaps. This method is used to detect weak sequence similarities between evolutionary distant sequences, in which deletions and insertions have occured over time. The classic example is *Smith-Waterman* local alignment, in which the two subsequences $\mathbf{a}, \mathbf{b}$, e.g. GATGC and GCTC, can be aligned as GATGC and GCT-C, i.e., with one gap (see Figure 2.4). The score function for alignment with gaps is given by

$$S[A] = \sum_{a,b} s_{a,b} - \delta N_g \,, \tag{2.164}$$

where N_g is the number of gaps with cost δ.

The *gapped alignment* of two sequences can be represented as a directed path on a two-dimensional lattice, see Figure 2.4. The alignment score is the sum over local scores of the traversed bonds, whereby diagonal bonds are gaps with penalty δ, while horizontal bonds are given the similarity scores $s(r,t) \equiv s_{a_i, b_j}$. What is sought for is the best scoring path connecting the lattice origin $(0,0)$ to its end, $(0, 2N)$.

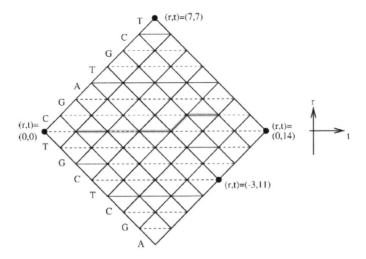

FIGURE I.2.4: The local alignment of the two sequences CGATGCT and TGCTCGA is represented as a directed path on an alignment lattice: the diagonal bonds correspond to gaps, while horizontal bonds are aligned pairs (Reprinted with permission from R. BUNDSCHUH, 2002. Copyright: American Physical Society.)

If we denote by $h(r,t)$ the score of the best path ending in a lattice point $h(r,t)$, the highest scoring *global alignment* can be computed with the *Needleman-Wunsch algorithm*

$$h(r,t+1) = \max\{h(r,t-1) + s(r,t), h(r+1,t) - \delta, h(r-1,t) - \delta\}. \quad (2.165)$$

This expression can be interpreted to describe the configuration of a *directed polymer* in a random potential given by the local scores $s(r,t)$. Another interpretation, which is somewhat easier to visualize, is to understand the configuration as the height profile $h(r,t)$ of a growing interface between a solid and, say, a liquid phase.

These systems - directed polymers in a random potential or a growing interface - are well-known in the physics of nonequilibrium systems. Due to the mapping of the sequence alignment problem onto the interface growth problem one can immediately show, from the knowledge of the models in nonequilibrium physics, that it belongs to the so-called KPZ *universality class*. A universality class comprises all model systems that can be shown to be mathematically equivalent and will have the same phase transition - i.e., they can be characterized by a common set of critical exponents.

Within a continuum description, one can show that the evolution of the heights is governed by the *Kardar-Parisi-Zhang equation* (often short: KPZ equation) (M. KARDAR, G. PARISI and Y.-C. ZHANG, 1986; T. HWA and M. LÄSSIG, 1998)

$$\partial_t h = \nu_0 \partial_r^2 h + \lambda_0 (\partial_r h)^2 + \eta(r, t) \tag{2.166}$$

where $\eta(r,t) = \frac{1}{2} s(r,t) - \delta$ is an uncorrelated Gaussian white noise. For $t \to \infty$ the heights assume a stationary equal-time distribution

$$P[h(r, t \to \infty)] \propto \exp\left(-\frac{1}{2D} \sum_r [h(r+1,t) - h(r,t)]^2\right). \tag{2.167}$$

The parameter D in this expression is a function of the scoring parameters μ and δ.

Typical score profiles $h(r,t)$ are illustrated in Figure 2.5 (top). In order to extract the characteristics of these profiles, it is instructive to consider the width of the profile, $w(t)$, defined by

$$w^2(t) = \frac{1}{X} \sum_{x=-X/2}^{x=X/2} [h(x,t) - h(t)]^2 \tag{2.168}$$

with the interval size $X (\approx N)$, and where the spatial average of the height, $h(t)$ is defined by

$$h(t) = \frac{1}{X} \sum_{x=-X/2}^{x=X/2} h(x,t). \tag{2.169}$$

The behaviour of h and w is shown in Figure 2.5 (bottom). As illustrated in the Figure, the asymptotic behaviour of the width obeys a scaling law

$$\overline{w}(t) = B(\mu, \delta) t^\omega \tag{2.170}$$

where the value of ω is known from the universal scaling behaviour of the KPZ equation. Note that the scaling behaviour is obtained from an ensemble average over different realizations of random sequences. The quantity

$$\Delta h(t) = h_{max}(t) - h_{min}(t) \tag{2.171}$$

displays the same scaling behaviour as w.

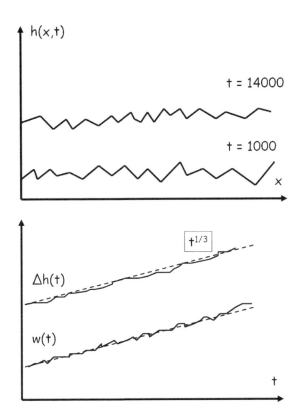

FIGURE I.2.5: Score profiles $h(x,t)$ (top) and scaling result (bottom) for $\Delta h(t)$ and $w(t)$ in a log-log plot (after T. Hwa and M. Lässig, 1998).

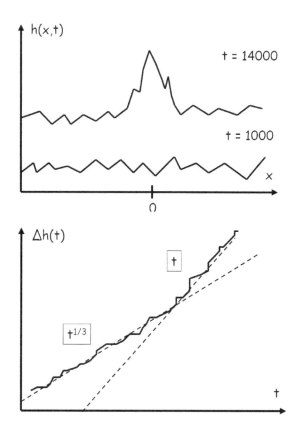

FIGURE I.2.6:   Score profiles $h(x,t)$ (top) for weakly correlated sequences, and scaling result for $\Delta h(t)$ in a log-log plot (after T. HWA and M. LÄSSIG, 1998).

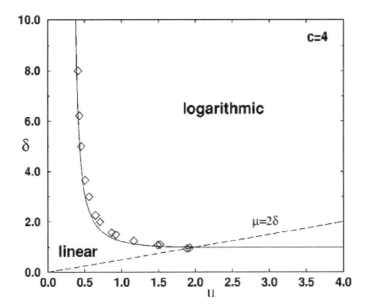

FIGURE I.2.7: The phase diagram of the log-linear phase transition in the parameters δ vs. μ. The variable c the number of letters in the alphabet; a value of 4 hence refers to the alphabet of DNA. (Reprinted with permission from R. BUNDSCHUH, 2002. Copyright: American Physical Society.)

If the sequences exhibit correlations, the behaviour changes. This is illustrated in Figure 2.6, which displays the growth of a peak in the score profiles. In terms of the height difference $\Delta h(t)$ one sees a deviation from the scaling behaviour of the random case, from a power law of the form $t^{1/3}$ to a linear law. This change in exponent illustrates the onset of global alignment, and we are thus back to the linear phase we discussed before for the case of gapless alignment.

As a final step we want to locate the logarithmic phase in parameter space. For the gapped case, a negative drift $\langle s \rangle$ is not sufficient, since the average score now has to grow by a gap-dependent amount $u(\{s_{a,b}\}, \delta)$ on top of the expectation value $\langle s \rangle$. Consequently, the log-linear transition occurs at

$$u(\{s_{a,b}\}, \delta) + \langle s \rangle = 0 \,. \tag{2.172}$$

In a (δ, μ) diagram this condition corresponds to a line $\delta_c(\mu)$, which can be calculated approximately or numerically. The result is shown in Figure 2.7.

Additional Notes

A basic reference for the description of nonequilibrium processes in physics and chemistry is the book by N. VAN KAMPEN, 1992, which served as a basis for the discussion presented here. Readers willing to learn more details are thus recommended to continue further studies there.

We have treated chemical kinetics in a very brief manner; for a much more detailed exposition, in particular on the kinetics of enzymatic reactions, readers are asked to consult the book by A. CORNISH-BOWDEN, M. JAMIN and V. SAKS, 2005.

The relation between equilibrium and nonequilibrium physics as expressed by the Jarzynski theorem is an active field of research. In particular the recent papers by C. JARZYNSKI, G. CROOKS, C. MAES and U. SEIFERT are recommended reading for those who want to know more on this topic - consultation of the current literature is, however, required.

On the experimental side, there are continuous efforts to validate the fluctuation theorems. Three recent contributions are by D. KELLER et al., 2003, E. H. TREPAGNIER et al., 2004, and D. COLLIN et al., 2005.

The literature on sequence alignment is (fairly obviously) extensive. A point of reference is the book by M. S. WATERMAN, 1995. The classic paper on alignment is by S. B. NEEDLEMAN AND C. D. WUNSCH, 1970. For a detailed description of the development until 1995 the reader is asked to consult the *Sources and Perspectives* section in Waterman's book.

The relationship between the alignment problem and models from nonequilibrium statistical mechanics was observed by T. HWA AND M. LÄSSIG, 1996 and 1998. More recently, a mapping to the asymmetric exclusion process has been proposed by R. BUNDSCHUH, 2002, which is the same universality class as the KPZ equation discussed in the text. The ASEP is a favorite model for studies of nonequilibrium statistical mechanics since it is amenable to rigorous approaches.

References

R. Bundschuh, *Asymmetric exclusion process and extremal statistics of random sequences*, Phys. Rev. E **65**, 031911 (2002)

D. Collin, F. Ritort, C. Jarzynski, S. B. Smith, I. Tinoco Jr. and C. Bustamante, *Verification of the Crooks fluctuation theorem and recovery of RNA folding free energies*, Nature **437**, 231-234 (2005)

A. Cornish-Bowden, M. Jamin and V. Saks, *Cinétique Enzymatique*, EDP Sciences (2005)

G. E. Crooks, *Path-ensemble averages in systems driven far from equilibrium*, Phys. Rev. E **61**, 2361-2366 (2000)

D. T. Gillespie, *Exact Stochastic Simulation of Coupled Chemical Reactions*, J. Phys. Chem. **81**, 2340-2361 (1977)

J. Gore, F. Ritort and C. Bustamante, *Bias and error in estimates of equilibrium free-energy differences from nonequilibrium measurements*, Proc. Natl. Acad. Sci. USA **100**, 12564-12569 (2003)

T. Hwa and M. Lässig, *Similarity Detection and Localization*, Phys. Rev. Lett. **76**, 2591-2594 (1996)

T. Hwa and M. Lässig, *Optimal Detection of Sequence Similarity by Local Alignment*, in *Proceedings of the Second Annual Conference on Computational Molecular Biology*, S. Istrail (ed.), ACM Press (1998)

C. Jarzynski, *Nonequilibrium Equality for Free Energy Differences*, Phys. Rev. Lett. **78**, 2690-2693 (1997)

C. Jarzynski, *Equilibrium free-energy differences from nonequilibrium measurements: A master-equation approach*, Phys. Rev. E **56**, 5018-5035 (1997)

M. Kardar, G. Parisi and Y.-C. Zhang, *Dynamic Scaling of Growing Interfaces*, Phys. Rev. Lett. **56**, 889-892 (1986)

D. Keller, D. Swigon and C. Bustamante, *Relating Single-Molecule Measurements to Thermodynamics*, Biophys. J. **84**, 733-738 (2003)

P. Langevin, *Sur la théorie du mouvement brownien*, C. R. Acad. Sci. (Paris) **146**, 530-533 (1908)

M. Lax, *Fluctuations from the Nonequilibrium Steady State*, Rev. Mod. Phys. **32**, 25-47 (1960)

J. Liphardt, S. Dumont, S. B. Smith, I. Tinoco Jr. and C. Bustamante, *Equilibrium Information from Nonequilibrium Measurements in an Experimental Test of Jarzynski's Equality*, Science **296**, 1832-1835 (2002)

C. Maes, *On the origin and use of fluctuation relations for the entropy*, Sém. Poincaré **2**, 29-62 (2003)

S. B. Needleman and C. D. Wunsch, *A general method applicable to the search for similarities in the amino acid sequence of two proteins*, J. Mol. Biol. **48**, 443-453 (1970)

S. Park, F. Khalili-Araghi, E. Tajkhorshid and K. Schulten, *Free energy calculation from steered molecular dynamics simulations using Jarzynski's e-quality*, J. Chem. Phys. **119**, 3559-3566 (2003)

U. Seifert, *Fluctuation theorem for birth-death or chemical master equations with time-dependent rates*, J. Phys. A: Math. Gen. **37**, L517-L521 (2004)

U. Seifert, *Fluctuation theorem for a single enzym or molecular motor*, Europhys. Lett. **70**, 36-41 (2005)

E. H. Trepagnier, C. Jarzynski, F. Ritort, G. E. Crooks, C. J. Bustamante and J. Liphardt, *Experimental test of Hatano and Sasa's nonequilibrium steady-state equality*, Proc. Natl. Acd. Sci. USA **101**, 15038-15041 (2004)

N. G. van Kampen, *Stochastic Processes in Physics and Chemistry*, Elsevier (1992)

M. Waterman, *Introduction to Computational Biology*, Chapman & Hall/ CRC (1995)

Part II

Biomolecules

Chapter 1

Molecules, Code and Representation

The proper term for such a translation rule is, strictly speaking, not a code but a cipher. [...] I did not know this at the time, which was fortunate because "genetic code" sounds a lot more intriguing than "genetic cipher".

Francis Crick, What Mad Pursuit (1988)

In this Chapter, we give a brief list of the basic chemical properties of the biomolecules whose physical properties will occupy us in the following chapters. The details given here are minimal; the indispensable reference to check is the book by B. ALBERTS et al. (2002).

1.1 DNA and RNA: the building blocks

DNA and *RNA* are charged polymers composed of three structural elements, a sugar, a phosphate group and the bases. The sugar gives the molecules their names: deoxyribose or ribose. Both are pentose rings with five carbon atoms; the arrangement of the carbons and their numbering is indicated in Figure 1.1. The chemical formula of deoxyribose is $C_5H_{10}O_4$, of ribose $C_5H_{10}O_5$. The sugar builds, together with the phosphates, the backbone of DNA and

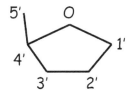

FIGURE II.1.1: The sugar ring of the backbone of both DNA and RNA. It contains an oxygen atom; the numbering of the carbon atoms is indicated.

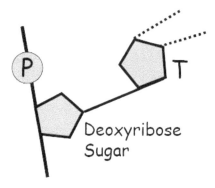

FIGURE II.1.2:   A DNA single strand with a thymine base offering two hydrogen bonds to bind to a complementary base, adenine.

RNA. The phosphates are attached to one sugar at the 5'-tail and to the next at the 3'-group, see Figure 1.2. Towards the opposite side, nucleotide bases are attached. Figure 1.3 displays the chemical structure of the four possible bases in DNA, two pyrimidines, and two purines, and the base uracil which replaces thymine in RNA. Neighbouring bases along a DNA strand experience *stacking interactions* when they are in registry, see Figure 1.4.

Consequently, a single-stranded DNA or RNA molecule is characterised by its base sequence and the orientation of the strand, e.g.

$$5' - ACTGTTTTACCCG - 3'.$$

The strand orientation is from the 5' to the 3' prime end; this is called the *sense strand*, while the direction 3' to 5' is the *anti-sense strand*. The bases can provide hydrogen bonds to a complementary base with which it can *hybridize* to form a double strand. This base pairing mechanism gives rise to the double-helical structure of the DNA molecule, see Figure 1.5.

In RNA, the base thymine (T) is replaced by uracil (U), and the sugar is ribose instead of 2-deoxyribose. In an organism RNA typically comes as a single strand, since it is the product of the *transcription* (i.e., the reading process) of a gene by the readout-molecule RNA polymerase. (We will learn more about those processes in Part III of the book.) The basic variants of RNA are listed in Table 1.1

While DNA usually forms a double-strand. between two complementary

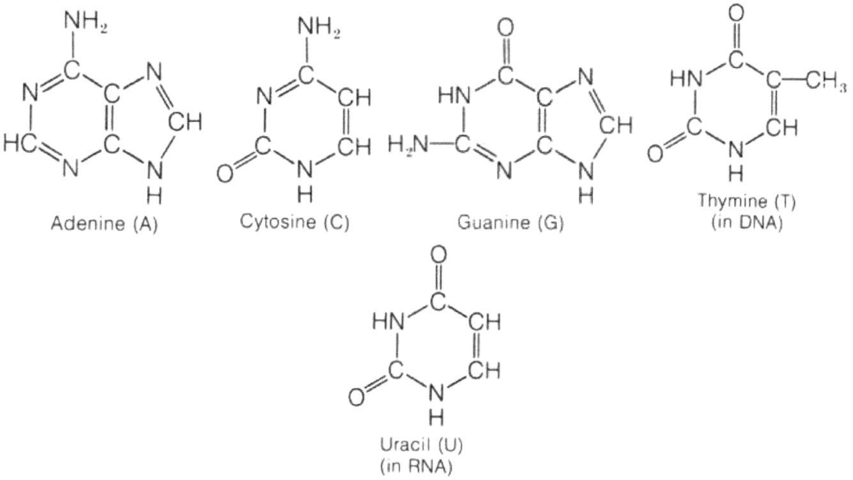

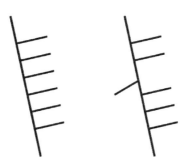

FIGURE II.1.3: DNA and RNA bases. Uracil replaces thymine in RNA.

FIGURE II.1.4: DNA and RNA stacking. Left: stacked bases, right: an unstacked base.

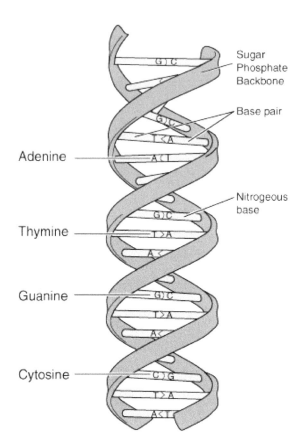

FIGURE II.1.5:   The DNA double helix. (From the Talking Glossary of Genetics, US National Human Genome Research Institute.)

sequences, RNA frequently hybridizes with itself.[1] As a consequence of *self-hybridization*, an RNA base sequence can give rise to a rich *secondary structure*, composed from RNA single strands interspersed with helical, i.e., bound, regions.

These different structures correspond to the different functional roles the specific RNA molecule can play within its biological context. The most variable type is the *messenger RNA* (mRNA) which is synthesized as a transcript from the coding regions of DNA; it is the read-out of the transcription process by RNA polymerase. The fold of a *transfer RNA* is illustrated in Figure

---

[1]DNA can also self-hybridize, but this is the less frequent situation.

1.6. As its name says, transfer RNA has a transport function; it brings an mRNA transcript to the cellular bodies where protein synthesis occurs, the ribosomes. It has a characteristic cloverleaf structure, in which each of the leaves takes up a particular function.

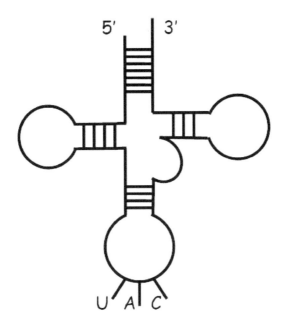

FIGURE II.1.6: Transfer RNA, in a schematic representation.

The original understanding of RNA as being a purely passive intermediate on the way from the genetic code of DNA to the functional properties of proteins has obviously evolved in the past years, in particular through the discovery of the mechanism of RNA interference, in which small non-coding *double-stranded RNA* molecules can regulate gene expression (A. FIRE et al., 1998). Apart from the classic variants, meanwhile a zoo of different RNA types has been found, some of which are indeed also double-stranded. RNA molecules are nowadays understood as being mostly non-coding, i.e., they do not help translate a DNA sequence into a protein. A list of the novel examples is provided in Table 1.2.

One could be tempted to say that DNA is all *information*; by contrast, RNA is *information and structure*. DNA structure can vary between different

TABLE 1.1: RNA variants

RNA type	size (nt)	function
mRNA	variable	transcript: template for protein
tRNA	75-95	adapter molecule
rRNA	$\approx 10^3$	part of ribosome; protein synthesis

TABLE 1.2: Recent RNA variants.

sRNA	small (non-coding) RNA, common name in bacteria, $75 - 400$ bp
ncRNA	non-coding RNA, common name in eukaryotes
miRNA	microRNA, form a putative translational regulatory gene family
stRNA	small temporal RNA, often also called miRNA
snRNA	small nuclear RNA, includes eukaryote spliceosomal RNA
snmRNA	small non-mRNA, ncRNA
snoRNA	small nucleolar RNA, involved in rRNA modification
tmRNA	bacterial ncRNA, have both mRNA and tRNA function
siRNA	short (21-25 nt) interfering RNA, double-stranded, involved in gene silencing

conformational and topological states, but these are usually based on a double-stranded molecule. By contrast, the single-stranded RNA molecule can build a *sequence-dependent* three-dimensional structure, a *fold*, and in this way it acquires similar degrees of freedom as a protein. It is thus *not sufficient* to know the base sequence in order to characterize RNA: one also needs to know the fold in order to understand its function.

1.2   Representing RNA structure

The basic structural elements that occur in RNA secondary structure are summarized in Figure 1.7. This collection already indicates the complexity one encounters when one wants to classify and predict these structures in the context of a complete molecule.

RNA structure is complex, but the number of the basic variable elements is only four, hence small when compared to the twenty *amino acids* of the proteins which we will encounter in next section. It is useful to briefly list the ideas used to represent the self-hybridized configurations of an RNA molecule. Two kinds of representations are shown in Figure 1.8.

The top graph of Figure 1.8 shows an RNA molecule represented explicitly by vertices (the nucleotides) and full-line edges. The edges are the connections between the vertices and represent the hydrogen bonds, shown as dotted lines; in the following, we drop this distinction. But there is also, different from Figure 1.7, a second class of edges, drawn in broken lines.

Consider the difference between the two configurations, once without and once with the broken edges. Without the broken edges, the RNA could be cut in two without affecting the hybridized chains. This is not the case anymore when the broken edges are present: the RNA structure now has a *pseudoknot*, intertwining separate paired regions along the chain.

This property becomes clear in the bottom graph. It can be thought to be obtained from the top graph by 'pulling at the ends' of the RNA chain. The *arc* or *rainbow diagram* clearly shows the 'overlapping' bonds for the pseudoknot.

There are formal ways to represent these RNA configurations which can be used in computer implementations (see, e.g., G. VERNIZZI et al., 2004). The first is a bracketing rule: an unpaired base is given by a dot, while a paired base is described by a bracket, indicating opening or closing of a bond. The top graph in Figure 1.7 would thus be represented as

$$.(((.....)))....(((.....))).$$ (1.1)

for the case without pseudoknot, and as

$$.(((...[)))....((((]...))).$$ (1.2)

when the pseudoknot-bonds are present. Each additional pseudoknot requires a new bracket-type to be introduced.

Alternatively, one can represent the bonds along the chain also in an array or matrix form. For this one writes the sequence of L nucleotides as an $L \times L$ *contact matrix* C with elements $C_{ij} = 1$, if i is paired to j, or zero otherwise. One can also interpret the pairing between any two bases i and j as a transposition of the elements $\{i, j\}$ and associate a permutation structure to it via $\sigma(i) = j$ if i, j are paired, and $\sigma(i) = i$ if not. For the example sequence $\{5' - CUUCAUCAGGAAAUGAC - 3'\}$ we give the pseudoknotted structure in dot-bracket notation in the first row and the permutation structure in the second and third:

$$\sigma = \begin{pmatrix} . & (& (& (& . & [& [& [&) &) &) & . & . &] &] &] & . \\ 1 & 2 & 3 & 4 & 5 & 6 & 7 & 8 & 9 & 10 & 11 & 12 & 13 & 14 & 15 & 16 & 17 \\ 1 & 11 & 10 & 9 & 5 & 16 & 15 & 14 & 4 & 3 & 2 & 12 & 13 & 8 & 7 & 6 & 17 \end{pmatrix}$$ (1.3)

This is an involution since σ^2 is the identity permutation.

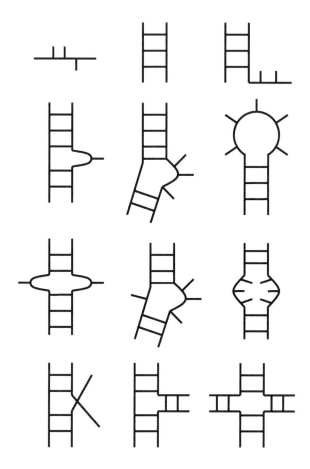

FIGURE II.1.7:   Schematic representation of the elements of RNA secondary structure; backbone and base pairs are both indicated as black bars, so no difference is made between covalent and hydrogen bonds. From top left to bottom right: single strand; duplex; duplex with dangling end; single nucleotide bulge; three nucleotide bulge; hairpin: stem (duplex region) and loop; mismatch pair or symmetrical loop; asymmetric internal loop; symmetric internal loop; two-stem junction; three-stem junction; four-stem junction.

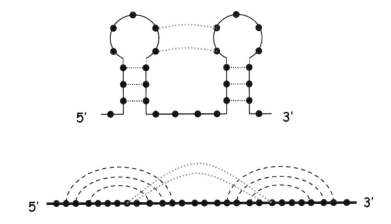

FIGURE II.1.8: Graphical representation of RNA with pseudoknots (see text).

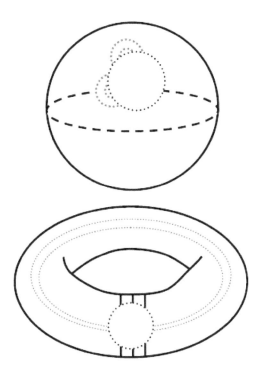

FIGURE II.1.9: Drawing RNA on a sphere and on a torus, after G. V-ERNIZZI et al., 2004.

Finally, there is an even more fundamental way to look at RNA *pseudo-knots*. If one closes the endpoints of a *rainbow diagram* to let it form a closed loop, one obtains a *circle diagram*. In this diagram, RNA structures with and without pseudoknots are distinguished by the presence or absence of edge crossings, when all of them are drawn either inside or outside the circle diagram.

Exercise. Draw the circle diagram for the kissing hairpin, with and without the pseudoknot bonds, and with bonds lying either all inside or outside the circle.

It is instructive to draw the circle diagram on the surface of a sphere, with the edges outside of the circle. Figure 1.9 (top) shows this for a non-pseudoknotted case. By contrast, a pseudoknotted configuration can be drawn without edge crossings on the surface of a torus (Figure 1.9, bottom).

The relationship between RNA structures and topology can be made quantitative through the *Euler characteristic* of the surface on which they can be drawn without crossing the bonds. The Euler characteristic is defined as

$$\chi \equiv V - E + F \tag{1.4}$$

where V are the vertices (nucleotides), E the edges (bonds), and F the faces of closed loops. RNA chains without pseudoknots have $\chi = 1$. For the kissing hairpin, one has $\chi = -1$. This result can also be expressed in terms of the *genus g* of the surface, which is $\chi = 1 - 2g$ for the case at hand. It is then clear that the kissing hairpin can be represented by a torus, for which $g = 1$. We will come back to the topological classification of RNA pseudoknots in Chapter 2 of this Part.

1.3   Proteins

Proteins are built-up from twenty *amino acids*; they are listed in Table 1.3 (following A. VON HAESELER and D. LIEBERS, 2003). A triplet of DNA bases codes for an amino acid. Since the number of existing amino acids is lower than the combinatorial possibility based on the triplet rule, there is thus redundancy in the code. It appears typically, but not exclusively, in the last codon. Note that particular combinations of the bases also code for the start and stop of coding sequences. Predominantly hydrophobic amino acids are marked with an asterisk; no distinction as to their degree of hydrophobicity is made.

TABLE 1.3:   Codons and amino acids; s.c.: starter codon.

Bases			Name	Abbreviations	
A	G	AG	arginine	Arg	R
A	G	UC	serine	Ser	S
A	A	AG	lysine	Lys	K
A	A	UC	asparagine	Asn	N
A	C	UCAG	threonine	Thr	T
A	U	G	methionine*/s.c.	Met	M
A	U	UCA	isoleucine*	Ile	I
C	G	UCAG	arginine	Arg	R
C	A	AG	glutamine	Gln	Q
C	A	UC	histine	His	H
C	C	UCAG	proline*	Pro	P
C	U	UCAG	leucine	Leu	L
U	G	G	tryptophane*	Trp	W
U	G	A	stop codon		
U	G	UC	cysteine*	Cys	C
U	A	G	stop codon		
U	A	A	stop codon		
U	A	UC	tyrosine*	Tyr	Y
U	C	UCAG	serine	Ser	S
U	U	AG	leucine*	Leu	L
U	U	UC	phenylalanine*	Phe	F
G	G	UCAG	glycine*	Gly	G
G	A	AG	glutamic acid	Glu	E
G	A	UC	aspartic acid	Asp	D
G	C	UCAG	alanine*	Ala	A
G	U	G	starter codon (s.c.)		
G	U	UCA	valine*	Val	V

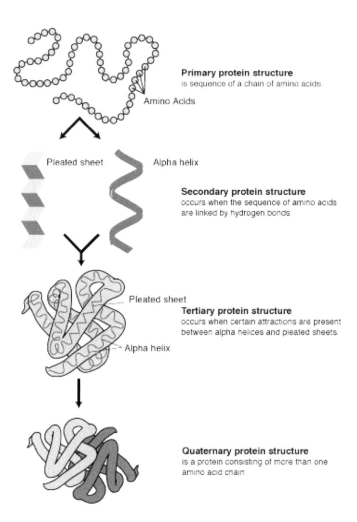

Primary protein structure
is sequence of a chain of amino acids

Amino Acids

Pleated sheet   Alpha helix

Secondary protein structure
occurs when the sequence of amino acids
are linked by hydrogen bonds

Pleated sheet

Tertiary protein structure
occurs when certain attractions are present
between alpha helices and pleated sheets

Alpha helix

Quaternary protein structure
is a protein consisting of more than one
amino acid chain

FIGURE II.1.10: Protein structure. (From the Talking Glossary of Genetics, US National Human Genome Research Institute.)

Amino acids can further be distinguished by their *hydrophobicity*: hydrophobic amino acids will prefer to bury themselves inside of a protein fold in order to avoid water contact.

The sequence of amino acids is called the *protein primary structure*. Proteins also form *secondary structures* which are recurrent structural elements. These are the *α-helix*, and the *β-sheet*; as the names indicate they correspond to a helical and a folded, but fairly planar structure. Like DNA, the molecular building elements form hydrogens bonds with each other to stabilize these structures. On the next level of complexity of proteins arises the full fold, the *tertiary structure*. Finally, fully folded proteins can form *protein complexes* or *quaternary structures*. These elements are all summarized in Figure 1.10.

References

B. Alberts, A. Johnson, J. Lewis, M. Raff, K. Roberts and P. Walter, *The Molecular Biology of the Cell*, 4th ed., Garland Science (2002)

A. Fire, D. Albertson, S. W. Harrison and D. G. Moerman, *Production of antisense RNA leads to effective and specific inhibition of gene expression in C. elegans muscle*, Development **113**, 503-514 (1991)

A. von Haeseler, D. Liebers, *Molekulare Evolution*, Fischer (2003)

G. Vernizzi, H. Orland and A. Zee, *Prediction of RNA pseudoknots by Monte Carlo simulations*, preprint q-bio/045014 (2004)

Chapter 2

DNA

We have stated in the first Chapter that the most fundamental feature of DNA is its ability to hybridize two complementary strands. This is the prime example of the physico-chemical recognition processes which are at the heart of essentially all biological mechanisms on a molecular level.

In some sense, the base-pairing mechanism of DNA is also the simplest of these recognition processes. It can be easily studied experimentally *in vitro*: one only needs to heat a solution containing double-stranded DNA molecules. The *thermal denaturation* or *melting* of DNA is the dissociation of the double-stranded DNA into its two separate single stands - hence just the inverse of the recognition mechanism. This process is reversible: by cooling the sample, the single-strands rehybridize to the double-strand. As first noticed by R. THOMAS in the early 1950's, this denaturation/hybridization process is indeed a thermodynamic phase transition in the limit of long chains. Many features of this phase transition remain correct in a more direct biological context, namely when the double strand of DNA is opened by the application of a direct force, e.g., by an enzyme.

We will now start to describe the thermal denaturation process, and we begin with short chains. This case is simpler than that of longer chains and also has important applications in biotechnology, e.g., for DNA microarrays to which we will turn later on as well.

2.1 Thermal stability of DNA: the melting transition

Denaturing short chains. The opening of double-stranded DNA molecules has two aspects. Ignoring first any sequence-dependent effects, it can be understood as a simple dissociation process: as one double strand opens, it gives rise to two single strands. One can easily imagine that when the chains become longer and longer, the sequence-dependence will become more and more important, and dissociation will not simply be one double strand vs. two single strands anymore; many intermediate states will become possible in which the double-stranded molecule is only partially dissociated.

As we have said, in the case of short chains (typically up to, say, 50 bp) we want to model the denaturation process as a dissociation equilibrium

$$C_2 \leftrightarrow 2C_1 \tag{2.1}$$

where C_1 is the concentration of single strands, and C_2 the concentration of double-strands. This reaction is governed by the equilibrium constant $K_D = C_2/C_1^2$. The total concentration of DNA is given by $C_T = C_1 + 2C_2$, since each double-strand contains two single strands. The quantities of interest are the fraction of double strands,

$$\theta_D \equiv \frac{2C_2}{C_T} . \tag{2.2}$$

or, likewise, the fraction of single strands, $\theta = 1 - \theta_D$. Using the definition of K_D, a quadratic equation for C_1 can be obtained which allows to express θ_D or θ in terms of K_D and C_T. The fraction of single strands is then found to fulfill the equation

$$\theta = \frac{-1 + \sqrt{1 + 2\gamma}}{\gamma} , \tag{2.3}$$

where $\gamma \equiv 4C_T/K_D$. This calculation assumes differing strands; if the two strands are self-complementary, $4C_T$ has to be replaced by C_T. *Why? Exercise!*

What remains to be calculated is the equilibrium constant K_D; in particular, its temperature dependence will turn out to be important. One has

$$K_D = \exp[-(\Delta H - T\Delta S)/RT] \tag{2.4}$$

where the transition enthalpy is given by

$$\Delta H = \Delta H_{H-bond} + \Delta H_{nn} \tag{2.5}$$

contains both contributions from the hydrogen bonds between the paired bases on the two strands, and from the *stacking* of the pairs on top of each other, see the previous section. Typically, for the stacking contribution it is assumed that it can be considered as a purely nearest-neighbor effect, but there are situations (e.g., in the presence of mismatches) when this is not sufficient.[1]

The transition entropy contribution in eq.(2.4) is typically approximated by $\Delta S = \Delta S_{bp}[N_{AT} + N_{CG}]$ where ΔS_{bp} is a temperature-independent value,

---

[1]The same philosophy applies to RNA-RNA and RNA-DNA-duplexes, but: the values of the stacking interactions differ in all these cases.

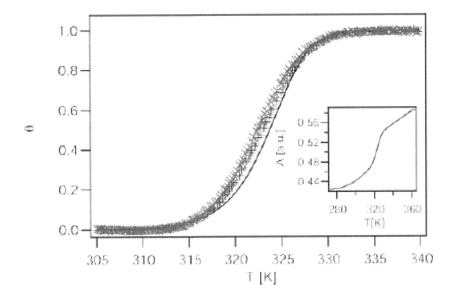

FIGURE II.2.1: Melting curve of a 16-bp oligomer, theory (line) and experiment. Insert: UV-absorption data. Reprinted with permission from J. BAYER et al.; Copyright (2005) American Chemical Society.

and N_{AT} and N_{GC} are the number of AT- and CG-base pairs, respectively. The precise values of the melting parameters have been determined experimentally, see, e.g., R. OWCZARZY et al., 1998, and J. SANTA-LUCIA, 1998.

Figure 2.1 illustrates the comparison between theory and experiment for the sequence 5'-TAG TTG TGA TGT ACA T-3'. (J. BAYER et al., 2005). The graph illustrates generic features observed in oligomer melting curves. The DNA double-strand undergoes a melting transition at a temperature T_M which is at $\theta = 0.5$; here, the slope of the curve is maximal (the derivative $d\theta/dT$ has a maximum). The melting temperature T_M depends on base composition and strand length. Further, the shape of the melting curve turns out to be universal, it generically has a sigmoidal form. The effect of a change in base sequence thus leads to a shift in melting temperature while the overall curve is unchanged. If the base sequence is extended by adding more bases, the melting temperature will shift to higher temperatures - simply because more bases have to be broken. At the same time, however, the sigmoidal curve will steepen up. This is a consequence of a *cooperative effect*: as long as the denaturation process occurs in a two-state fashion, essentially all bases have to open up in concert.

Melting long chains: the Poland-Scheraga model. We now turn to the discussion of melting of long DNA chains which do not open in the simple two-state fashion. We can easily imagine that if we make the chains longer and longer, from 50 bp to, say, 500 bp, the cooperative effect of all bases opening in concert will not be operative any more all along the chain. We may then expect that a DNA molecule undergoing the denaturation process may look like the schematic illustration in Figure 2.2. The molecule consists of a sequence of helices and open segments, so-called *denaturation loops* or *bubbles*. The configuration shown in addition has open ends, but closed ends are of course possible, too.

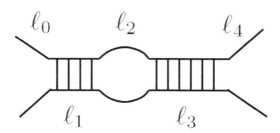

FIGURE II.2.2:   A DNA chain configuration near the melting transition: it consists of an open end of length ℓ_0, a bound segment of length ℓ_1, a loop of length ℓ_2, a bound segment of length ℓ_3, and an open end of length ℓ_4.

A model to describe the statistics of these configurations was first proposed by POLAND and SCHERAGA, 1970; we here follow the discussion by Y. KAFRI et al., 2002. If we ignore the configurational entropy of a bound segment embedded in the ambient space, its statistical weight will be given by

$$w^\ell = exp(-\ell E_0/k_B T) \tag{2.6}$$

where ℓ is the length of the segment, and E_0 is a binding energy in which, for the moment, no distinction is made between the chemical nature of the bases.

By contrast, a *denaturation loop* has no energy associated with it; its statistical weight is consequently determined only by its degeneracy - we have to count the number of its configurations in order to estimate its contribution to the entropy of the configuration.

Assuming that the DNA loop is fully flexible, in the simplest modelling it can be considered as a random walk which returns to its origin after a path of length 2ℓ. From this modelling idea one then knows that the statistical weight for the loop of length ℓ has an algebraic form,

$$\Omega(2\ell) = \sigma \frac{s^\ell}{\ell^c} \tag{2.7}$$

where s is some constant; the amplitude prefactor σ will be simply put to $\sigma = 1$ for the moment; but we will come back to it later.

The exponent c is determined by the statistical properties of the loop config-urations. Finally, the configuration of the chain ends must be characterized; they consist of two denatured strands of length ℓ with a weight $\Lambda(2\ell)$ of a similar form as in eq.(2.7), but with a different exponent $\bar{c}$. The values c and $\bar{c}$ can assume will be given later.

Given these weights, we can now proceed to calculate the total weight of any given configuration. Supposing the example shown in Figure 2.2, its statistical weight is obtained by

$$\Lambda(2\ell_0) w^{\ell_1} \Omega(2\ell_2) w^{\ell_3} \Lambda(2\ell_4). \tag{2.8}$$

The melting phase transition. This section is somewhat more formal since we now want to determine the nature of the transition between a bound and a denatured phase of a double-stranded DNA molecule.

In order to determine the thermodynamic properties of this model, it is practical for the calculation to assume a grand canonical ensemble in which the total chain length L is allowed to fluctuate.[2] The reason why this is a good choice is that in this ensemble the total partition function can be conveniently expressed as a geometric series,

$$\mathcal{Z} = \sum_{L=0}^{\infty} Z(L) z^L = \frac{V_0(z)Q(z)}{1 - U(z)V(z)}, \tag{2.9}$$

where $Z(L)$ is the canonical partition function of a chain of given length L, and z is the *fugacity*.[3] The functions $U(z)$, $V(z)$ and $Q(z)$ are defined as

$$U(z) \equiv \sum_{\ell=1}^{\infty} \Omega(2\ell) z^\ell = \sum_{\ell=1}^{\infty} \frac{(sz)^\ell}{\ell^c} = \Phi_c(sz), \tag{2.10}$$

---

[2] Remember that we argued in Part I, Chapter 1 that in the thermodynamic limit the ensembles are equivalent.

[3] The fugacity is the intensive variable conjugate to the chain length, as much as the chemical potential is conjugate to the particle number.

$$V(z) \equiv \sum_{\ell=1}^{\infty} w^\ell z^\ell = \frac{wz}{1-wz}, \tag{2.11}$$

$$Q(z) \equiv 1 + \sum_{\ell=1}^{\infty} \Lambda(2\ell) z^\ell = 1 + \Phi_{\bar{c}}(sz) \tag{2.12}$$

and

$$V_0(z) = 1 + V(z). \tag{2.13}$$

$\Phi_c(sz)$ is the *polylog function* which converges for $|z| < 1$; if in addition $\mathrm{Re}(c) > 0$, the function has an integral representation

$$\Phi_c(z) = \frac{1}{\Gamma(c)} \int_0^\infty dt\, t^{c-1} \frac{ze^{-t}}{1 - ze^{-t}}, \tag{2.14}$$

where $\Gamma(c)$ is Euler's gamma function. The integral representation allows to see that $\Phi_c(z)$ has a divergence of the form $|z - 1|^{c-1}$ for $z \to 1$, if $c \leq 1$. If $c > 1$ and $1 - z = \varepsilon \ll 1$, one has $\Phi_c(1) - \Phi_c(1-\varepsilon) \sim \varepsilon^\zeta$ where $\zeta = \min(1, c-1)$.

In order to fix the average chain length L one has to choose the fugacity such that

$$L = \frac{\partial \ln \mathcal{Z}}{\partial \ln z}. \tag{2.15}$$

The thermodynamic limit $L \to \infty$ is obtained by allowing z to approach the lowest fugacity value z^* for which the partition function $\mathcal{Z}$ diverges. This divergence can have two sources: either the numerator grows without bounds, or the denominator vanishes. In fact, both cases occur. At low temperatures, the denominator vanishes, which is at

$$U(z^*)V(z^*) = 1. \tag{2.16}$$

Making use of expression eq.(2.11) for $V(z)$, this result can be expressed as

$$U(z^*) = \Phi_c(sz^*) = \frac{1}{wz^*} - 1 = \frac{1}{V(z^*)}. \tag{2.17}$$

The solutions of this equation depend on the singularities of $\Phi_c(z)$, which depend itself on the value of c.

We take as the order parameter for the denaturation transition in the long chains the fraction of bound monomers[4], θ_b. Its temperature dependence in

---

[4] For the case of short chains, $\theta_b = \theta_D$.

the thermodynamic limit can be calculated from the behaviour of $z^*(w)$, since the average number of bound pairs in a chain is given by

$$\langle m \rangle = \frac{\partial \ln \mathcal{Z}}{\partial w} \tag{2.18}$$

so that

$$\theta_b = \lim_{L \to \infty} \frac{\langle m \rangle}{L} = \frac{\partial \ln z^*}{\partial \ln w}. \tag{2.19}$$

In the case of the short chains, we defined the transition as the point where the temperature-derivative of the fraction of bound (or, unbound) base pairs would show a maximum. We also noted that for somewhat longer chains, this curve would steepen, and we interpreted this as a cooperative effect. Indeed, it is a signature of a collective phenomenon, which in the limit of infinite system becomes *sharp*, provided all bases were to open up collectively.

If we now look at the *order parameters* $\langle m \rangle$ and θ_b we defined for the state of the DNA double-strand, we are asked to check what the possible limiting behaviour of these quantities in the limit $L \to \infty$ is. This whole discussion resembles the (simpler) discussion of the phase transition in the 1-dimensional Ising model in Part I of the book. The difference between the two is that, apart from the energy of the configurations, we have to explicitly account for the loop entropy. And that, as we will see, makes everything different since the entropic weights of the loops are algebraic in nature, hence of a long-range nature and not rapidly decaying as for an exponential contribution.

As a consequence of this observation, the critical quantity in this discussion is the exponent c. Depending on its value we can find three different scenarios:

- $c \leq 1$: **no phase transition.** The function $U(z)$ is finite for all $z < 1/s$, and it diverges at $z = 1/s$. The function $1/V(z)$ is always finite and intersects $U(z)$ for $z < 1/s$. There is no singular behaviour.

- $1 < c \leq 2$: **continuous phase transition.** In this case, $U(z)$ is finite at $z = 1/s$ since $c > 1$. For $z > 1/s$, it is infinite. The singular point $z_M = 1/s$ is thus the phase transition point, separating a bound and a denatured regime. At the transition, the derivative of $U(z)$ diverges.

- $c > 2$: **first-order phase transition.** In this case, $U(z)$ and its derivative are finite at $z = z_M$. Again, there is a transition at this point; above the transition, θ_b vanishes in the thermodynamic limit. The transition is thus first-order.

All that remains to do now is to fix the value of c, which so far has been left unspecified. Since we have modeled the loops as random walks, we have to quantify the relationship between random walks and the polymeric nature of DNA.

Indeed, if we were to consider the DNA loop simply as a random walk of a given length which returns to the origin, we would find a value of $c = d/2$ in d space dimensions. Taking this exponent, and looking into the list of possible behaviours, there would be no denaturation transition in $d \leq 2$; for $2 < d \leq 4$ the transition is continuous, and for $d > 4$, the transition will be first-order.

This argument can be somewhat refined when considering the loop as a *self-avoiding walk*, since for a purely random walk configurations can occur in which the walk crosses itself. The assumption of self-avoidance seems more realistic since a real polymer can of course not intersect itself. One has

$$c = d\nu \qquad (2.20)$$

where ν is the exponent associated with the *radius of gyration*[5] R_G of a self-avoiding random walk. For a walk of length L one has $R_G \sim L^\nu$ with $\nu = 3/4$ in $d = 2$ and $\nu \approx 0.588$ in $d = 3$. For the loop exponent c, this yields $c = 1.5$ in $d = 2$, and $c = 1.766$ in $d = 3$. The inclusion of self-avoidance thus leads to a slight smoothing of the transition.

All these reasonings are based on the assumption that the loop can be considered an isolated object which does not interact with the rest of the chain. In fact, the whole build-up of the weights assumes that one can consider the 'bits and pieces', i.e., the helices and loops, as essentially independent from each other. Recently, however, arguments have been put forward to account for the self-avoidance between the loops and the rest of the chain, albeit in an approximate way. These so-called *excluded-volume effects* between a loop and the chain have originally been derived using the theory of polymer networks by B. DUPLANTIER, 1986. The discussion of his theory goes beyond the scope of this book since it heavily relies on renormalization group arguments; here, it must suffice us to simply state that the excluded volume interactions modify the self-avoidance exponent relation eq.(2.20) into the expression

$$c = d\nu - 2\nu_3 \qquad (2.21)$$

where ν_3 takes into account the contribution from the two 3-vertices at the extremities of a loop, which connect the loop to the chain. In $d = 3$, one finds a value of $\nu_3 \approx -0.175$. Thus for c results

$$c \approx 2.115 \qquad (2.22)$$

i.e., a value which is slightly *larger* than two. This result is in accord with independent numerical work based on Monte-Carlo calculations for lattice

---

[5]The radius of gyration is the average distance of a monomer from the centre of mass of the polymer.

models of DNA (E. CARLON et al., 2002).

A side remark on the boundary effects. Just in order to complete the discussion, it turns out that the result on the denaturation transition does not depend on the value of $\bar{c}$. This exponent enters the discussion, e.g., if one is interested in the average length of the end segment near the transition, which is given by

$$\xi = z \frac{\partial \ln Q}{\partial z}\bigg|_{z=z^*} \tag{2.23}$$

with the value of z^* determined before. The three different behaviours found for the denaturation transition can be found back for the average end segment length; for more details of this behaviour, see Y. KAFRI et al., 2002. Here we simply note that the value of $\bar{c}$ can be obtained by summing up the scaling dimensions of a linear chain and a fork, leading to $\bar{c} = -(\nu_1 + \nu_3) \approx 0.092$, i.e., a very small value. Near the melting transition the end segment diverges according to

$$\xi \sim \frac{1}{|T - T_M|}, \tag{2.24}$$

where T_M is the melting temperature. This completes this somewhat special-istic discussion.

The melting transition with sequence dependence. The main result of the previous section was that the denaturation transition may in fact be a first-order rather than a continuous transition if excluded-volume effects be-tween the bubbles and the chain do matter.

This finding has revived a dormant, but long-ongoing controversy on the nature of the transition. We will now turn to the computation of the melting curves of specific DNA sequences and want to see whether the computational results can faithfully represent experimental data. We thus first have to build in sequence-dependence into the theory which we had ignored so far.

In order to systematically build in sequence effects we will follow a version of the Poland-Scheraga model which is also useful for further generalizations (like the thermal stability of hybrid DNA, i.e., chains which are not fully com-plementary and/or of different length). T. GAREL and H. ORLAND, 2004, have recently cast the Poland-Scheraga model into a recursive formulation based on the partition function.[6] Previous work by D. POLAND, 1974, had relied on a recursion formulation for configuration probabilities. Based on these, the simulation program MELTSIM had been developed by R. BLAKE

---

[6]We have seen a basic version of this procedure in the section on the Ising model in Part I of this book, and will encounter another application in the discussion of RNA secondary structure in the following Chapter.

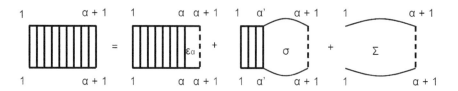

FIGURE II.2.3: Graphical representation of the Poland recursion for the partition function

et al., 1999, a variant of which had been used to obtain the simulation results described here.

In the original Poland-Scheraga model one only considers two complementary strands of equal length N. Let us call $Z_f(\alpha)$ the *forward partition function* of the two strands starting at base 1, and ending at base α (which is assumed paired). The interaction of the bases is built as that between base pair stacks at positions α and $\alpha + 1$ along the chain, with associated stacking energies[7]

$$\epsilon_\alpha \equiv \epsilon_{\alpha,\alpha+1;\alpha,\alpha+1} \, . \tag{2.25}$$

In order to find the recursion relation obeyed by $Z_f(\alpha + 1)$ one has to consider the three possibilities to pair two bases at position $\alpha + 1$:

- the last pair $(\alpha, \alpha + 1)$ is stacked;

- there is a loop starting at any α' with $1 \leq \alpha' \leq \alpha - 1$ which ends at $\alpha + 1$;

- there is no loop.

This is illustrated in Figure 2.3. Formally, this recursion relation is expressed as

$$Z_f(\alpha + 1) = e^{-\beta\epsilon_\alpha} Z_f(\alpha) + \sigma \sum_{\alpha'=1}^{\alpha-1} Z_f(\alpha') \mathcal{N}(2(\alpha + 1 - \alpha')) + \Sigma\mathcal{M}(\alpha) \tag{2.26}$$

where $\beta = 1/kT$. There are two *cooperativity parameters* in this expression: σ and Σ; these parameters quantify the probabilities for loop and fork formation, respectively, and are assumed to be sequence-independent.

---

[7]The values for these energy parameters are defined such as to include the hydrogen bond contributions.

The factor $\mathcal{N}$ counts the number of conformations of a chain beginning at base α' and ending at $\alpha + 1$. Asymptotically, one has

$$\mathcal{N}(2(\alpha + 1 - \alpha')) = \mu^{\alpha - \alpha'} f(\alpha - \alpha') \tag{2.27}$$

where $k_B \ln \mu$ is the entropy per base pair (irrespective of its nature), and $f(\ell)$ is the return probability of a loop of length 2ℓ to the origin. Finally, in eq.(2.26), $\mathcal{M}(\alpha) = \mu^{\alpha} g(\alpha)$, which counts the number of conformations of a pair of unbound chains starting at base 1 and paired at base $\alpha + 1$. The function g has a power-law behaviour, however, as we have seen before, with an exponent close to 0. We therefore simply put this factor to 1.

In a similar fashion as for the forward partition function Z_f one can compute the backward partition function Z_b, starting at base N and ending at a paired base α. Again there are three options to pair a base at position α, and thus one finds

$$Z_b(\alpha + 1) = e^{-\beta \epsilon_{\alpha}} Z_b(\alpha + 1) + \sigma \sum_{\alpha' = \alpha + 2}^{N} Z_b(\alpha') \mathcal{N}(2(\alpha' - \alpha)) + \Sigma \mathcal{M}(N - \alpha). \tag{2.28}$$

From the expressions eq.(2.26,2.28) one obtains the probability for the binding of a base pair α as

$$p_\alpha = \frac{Z_f(\alpha) Z_b(\alpha)}{Z}, \tag{2.29}$$

where Z is the partition function of the two strands, given by

$$Z = Z_f(N) + \Sigma \left(\mu Z_f(N - 1) + \mu^2 Z_f(N - 2) + ... + \mu^{N-1} Z_f(1) \right) \tag{2.30}$$

or, expressed in terms of the backward partition function,

$$Z = Z_b(1) + \Sigma \left(\mu Z_b(2) + ... \mu^{N-1} Z_b(N) \right). \tag{2.31}$$

For the implementation of these recursion relations one has to take into account that the algorithm is $\mathcal{O}(N^2)$ since one has to compute $\mathcal{O}(\alpha)$ terms for each value α. In order to reduce the computational complexity, M. FIXMAN and J. J. FREIRE, 1977, have developed an approximation in which the power-law loop-entropy factor is replaced by a sum of exponentials

$$f(\ell) = \frac{1}{\ell^c} \approx \sum_{i=1}^{I} a_i e^{-b_i \ell} \tag{2.32}$$

where the I parameters (a_i, b_i) obey a set of non-linear equations and determine the degree of accuracy. With this step, computational complexity is reduced to $\mathcal{O}(N \cdot I)$. In the case of sequences of unequal length, the complexity of the algorithm can reduced with this method from $O(N_1^2 N_2^2)$ to $O(N_1 N_2)$.

We can turn to the application of the PS-model, and we will see that the story that results will be, to some extent, a story of the cooperativity parameter σ.

2.2   The melting profiles of genomic DNA and cDNA

Melting genomic DNA. We now want to take a look at the comparison of the theory we have described with experimental data. In particular we ask what value of c, the one without excluded-volume or with inclusion of excluded-volume effects does fit experiments best? Could we even find out this way whether the transition is first or second order?

In order to perform such a quantitative comparison, the theory has finally to be complemented by experimental parameters. The first set of parameters are the nearest and paired neighbor energies (R. D. BLAKE et al., 1999), distinguishing the different paired bases and their stacks; there are ten independent values to be specified. Further, there is the amplitude factor, the *cooperativity parameter* σ which we had put to a value of one before. On a technical level, this parameter determines the relative magnitude of energetic and entropic effects. Its name, however, indicates its physical meaning: it determines how many bases interact cooperatively, i.e., open up together to disrupt the helix and form a loop.

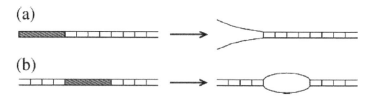

FIGURE II.2.4:   Schematic diagram of the linearization of an AT-insertion in a GC-rich bacterial plasmid. The bar represents the inserted DNA segment which upon cutting is either placed at an extremity of the molecule (a) or at its center (b). Reprinted with permission from R. BLOSSEY and E. CARLON; Copyright (2003) American Physical Society.

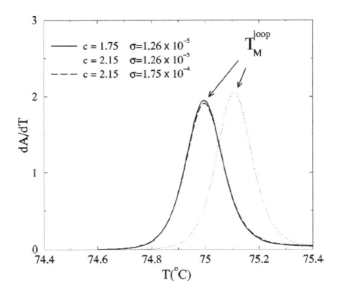

FIGURE II.2.5: Fits of the melting curve with different values of c and σ. The curve with $c = 2.15$ and $\sigma = 1.26 \times 10^{-5}$ does not fit the experimental data; by contrast, two choices for c and σ lie exactly on top of each other. Reprinted with permission from R. BLOSSEY and E. CARLON; Copyright (2003) American Physical Society.

A good system to test the issue of the order of the denaturation transition has been introduced by R. D. BLAKE and S. G. DELCOURT, 1998. They inserted artificial AT-rich sequences of varying length into GC-rich bacterial (plasmid) DNA, with sequence lengths between about 200 and 700 bp. After the insertion, the circular plasmid DNA was linearized in two ways: either the insertion was left at the extremity, or left imbedded in the GC-rich chain, see Figure 2.4. Both configurations differ in their melting temperatures: the sequence which melts from the extremity has a lower T_M, i.e., $T_M^{loop} > T_M^{end}$.

Figure 2.5 shows a first example of a calculated differential melting curve for the case of an embedded loop, for the longest inserted sequence with 747 bp, and for three sets of values (σ, c). It is found that the curves with the sets $(\sigma = 1.26 \times 10^{-5}, c = 1.75)$ and $(\sigma = 1.75 \times 10^{-4}, c = 2.15)$ fall on top of each other; the curve with the value $\sigma = 1.26 \times 10^{-4}$ and $c = 2.15$ has a higher T_M. Note that the two curves which fall on top of each other in this graph are in accord with experiment (data not shown). This result shows that a change in the value of c can apparently be compensated for by choosing a smaller value of σ.

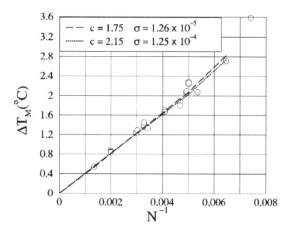

FIGURE II.2.6: Length dependence of the melting temperature shift. Reprinted with permission from R. BLOSSEY and E. CARLON; Copyright (2003) American Physical Society.

Figure 2.6 shows the theoretical values for the melting temperature difference $\Delta T_M \equiv T_M^{loop} - T_M^{end}$ the two sets (σ, c), in comparison with the experimental data, as a function of inverse insert length, $1/N$. Both theoretical curves deviate for shorter chain lengths, for which the theory is less reliable (see below). Given that the experimental resolution of melting temperatures is on the order of $0.1°$ C, there is obviously no direct way to decide the issue of the nature of the transition based on a comparison of the Poland-Scheraga model with this kind of experiment.

If one considers longer sequences of genomic DNA, such as, e.g., the human cancer-related gene eIF-4G with a sequence length of about 2900 bp, a structure-rich denaturation profile appears with many interior openings, hence loops, along the sequence. This is illustrated in Figure 2.7, where the differential melting signal is shown together with the denaturation probability $1 - A(i)$, which is the probability that the i-th base pair is in a closed state. The figure shows this probability at six different temperatures, labelled with greek letters $\alpha, .., \phi$. Again, in this Figure, the melting curves are compared for different values of σ and c, and as before for the plasmid inserts, again the curves can be recovered with two sets of parameter (and hence, consequently, in fact by a whole range of parameter values, interpolating between the two chosen values of σ and c).

One may wonder whether the whole discussion misses an important physical parameter of DNA, the stiffness of the double helix, which we encountered in Part I in the discussion of the WLC model. Since the persistence length

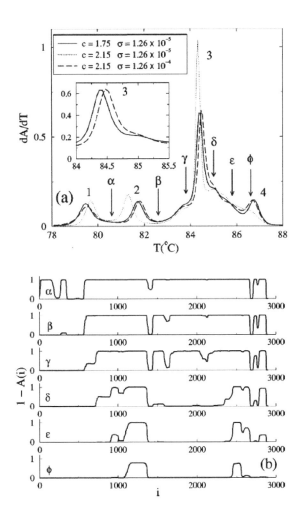

FIGURE II.2.7: Melting curve of the cancer-related human gene eIF-4G with about 2900 bp (top); opening regions along the chains at different temperatures (bottom). Reprinted with permission from R. BLOSSEY and E. CARLON; Copyright (2003) American Physical Society.

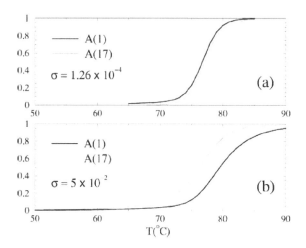

FIGURE II.2.8: Effect of a change of the cooperativity parameter on the opening probability of a short chain designed for loop opening. Reprinted with permission from R. BLOSSEY and E. CARLON; Copyright (2003) American Physical Society.

of the double-stranded DNA, ℓ_P^{ds}, is typically much larger than that of single-stranded DNA, the relevance of excluded-volume effects between the loop and the chain has been questioned (A. HANKE and R. METZLER, 2003). In their view, the double-stranded DNA enclosing the bubble is so stiff compared to the flexible single strands of the bubble that the excluded volume effect plays no role.

Several factors, however, intervene in this issue. Firstly, denaturation is a high-temperature phenomenon, since melting temperatures typically lie in the range between 60°C-90 °C, and not at room or physiological temperature, where values of the persistence length are often considered. Secondly, in the limit of high salt concentrations, electrostatics is fully screened and the dependence of ℓ_P on temperature can be assumed to follow the worm-like chain model where $\ell_P \sim T^{-1}$ (see Part I, Chapter 1).

A further effect affecting the value of the persistence length is the presence of small bubbles along the melting chain. It should be noted that the Poland-Scheraga model is tuned to describe long chains and the opening of long loops along the chain; the length of helix and loop segments should scale as the inverse cooperativity parameter, hence as $1/\sigma$. One could thus be misled to assume that short loops would be suppressed during the melting process, which is wrong.

This fact is illustrated in Figure 2.8, which plots the probability of finding a specific base pair, $i = 1$ and $i = 17$, in an open state in a chain of 30 bp length. Both $A(1)$ and $A(17)$ can be measured by fluorescence techniques; the base pair $i = 17$ is placed in the interior of an AT-rich region. The shortness of the sequence leads to the insensitivity of the melting curve on the value of c, but it sensitively depends on σ. For the small value of $\sigma \sim 10^{-4}$, both probabilities are indistinguishable, which would mean that there is no loop opening, and the sequence rather melts from the ends. Increasing the value of σ by a factor of 100 allows the opening of loops, and brings the theoretical result close to what is observed in experiment (G. ALTAN-BONNET et al., 2003). Consequently, it appears that $\sigma = \sigma(L)$.

We thus conclude that

- The Poland-Scheraga can be used to describe even complex denaturation profiles in quantitative accord with experiment;

- The theoretical results depend on the value of the critical exponent c and the cooperativity parameter σ, both need to be (and can be) adjusted to fit to experiment;

- The fit to experiment depends on the length of the DNA under scrutiny; by construction, the Poland-Scheraga model is more tuned to DNA properties on longer length scales;

- We cannot clarify the issue of the order of the phase transition; from a practical point of view for the melting profiles of DNA this seems almost an academic question. The transition is borderline between first and second order, beyond experimental resolution.[8]

Melting cDNA. DNA contains various sequence regions which are biological entities: first there are the *genes*, or more specifically, the protein-coding regions, the *exons*, but also the so-called junk DNA, non-coding regions, the *introns*. There are also *regulatory regions*: places where regulatory proteins attached and control the expression of genes (see Part III). If we melt purely genomic DNA, we will get a bit of everything. Can we distinguish between the melting behaviour of different sequence regions - is there a correlation between the biological purpose of a sequence and its (thermal) stability ?[9]

---

[8]Physicists protesting against this pessimistic conclusion should look back into the critical properties of classical superconductors.

[9]One might object that the thermal stability of DNA cannot be relevant for organisms since the melting temperatures of DNA are usually much higher than physiologically relevant temperatures. One counter argument against this is that a position which is easier to destabilize thermally can likewise be more readily destabilized either by changes of chemical conditions (pH) or the localized action of forces, e.g., by protein complexes on DNA.

We address this question for *complementary DNA* (cDNA).

Complementary DNA is a molecule which contains only exons (including so-called untranslated end regions, UTR's). Complementary DNA can be obtained from genomic DNA by first transcribing it into RNA, splicing out of the introns, and transcribing the mature RNA back into DNA; this latter step is done by a viral enzyme, the reverse transcriptase. Figure 2.9 shows the build-up of genomic DNA and cDNA in a schematic comparison.

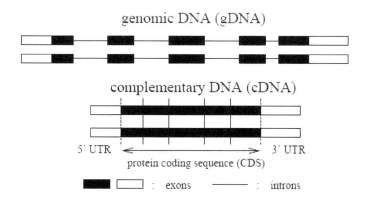

FIGURE II.2.9: Structure of genomic DNA vs. cDNA. Reprinted with permission from E. CARLON et al.; Copyright (2005) American Physical Society.

Melting of genomic DNA of various organisms has been studied by E. YERA-MIAN, 2000. Quite generally it is found that the melting curves reflect the base composition; since exons are on average more GC-rich than introns, which have a bias to AT bases, this difference can in some cases permit to distinguish genes and non-coding regions from each other. AT-rich regions melt more easily since the base pairing has only two hydrogen bonds, instead of the three between G and C. However, this distinction does not always work; e.g., for bacteria it was found that the structure of their ultra-dense genomes without non-coding regions does not permit to identify genes based on melting signatures alone.

This is indeed different for cDNA (E. CARLON et al., 2005). If one melts cDNA, one finds a differential melting curve similar to the one shown in Figure 2.10. As for genomic DNA, several distinct melting peaks arise. In the figure, sequence information has been added: the horizontal bars indicate the location of the boundaries between two exons as known from sequence annotation. The numbers on the right indicate the number of base pairs of the intron removed at that position. The vertical bars in the figure indicate when the probability of having bound pairs falls below 1/2. The ends of these bars locate the position of a *thermal boundary*: a sequence position at which the opening of the sequence is stopped at increasing temperature.

Figure 2.11 shows that in several cases thermal boundaries coincide with the annotated exon-exon boundaries, i.e., known coding regions of the genes. This observation holds for many human genes, with a resulting coincidence of thermodynamic and annotated boundaries at about 30 %.

At present, there is no simple explanation for this finding; but one may speculate. In this context it is useful to think in evolutionary terms. How did exons and introns come into their positions in the first place? There are two current, opposing hypotheses on the evolution of genomes, one based on the idea that exon positioning came late, and were inserted (shuffled around) into intronic DNA. (Bacteria have no introns: but according to the theory they were just very active in getting rid of the junk.) The opposing hypothesis considers exons to be evolutionarily 'early', and introns being inserted into an otherwise largely exonic DNA (W. GILBERT, 1987).

The finding that exon-exon-boundaries are, in humans, to about 30 % located in positions at which DNA is less (thermally) stable might support the view that introns are 'late': the location of thermal boundaries might be preferred sites for intron insertion, since in these positions double-stranded DNA is more readily opened up. But then, these spots might also be prone to intron losses. In any case, an agreement between positions at which a physical signature coincides with a biological signature to such an extent seems hard to be just accidental.

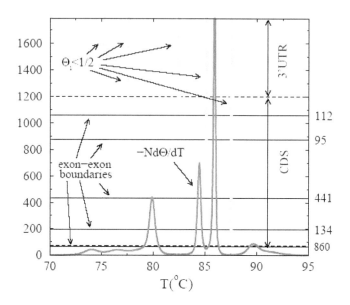

FIGURE II.2.10: Melting curve and exon-exon boundaries. For the explanation, see text; N is sequence length. Reprinted with permission from E. CARLON et al.; Copyright (2005) American Physical Society.

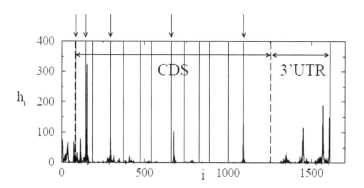

FIGURE II.2.11: Correlation of thermal boundaries with exon-exon boundaries. Reprinted with permission from E. CARLON et al.; Copyright (2005) American Physical Society.

2.3 Hybridizing DNA

The final section of the Chapter on DNA melting is devoted to a discussion of the hybridization of oligomeric DNA on microarrays in order to measure gene expression levels. The idea of this method is that the amount of RNA in a cell (or, of back-transcribed RNA into cDNA) can serve as a measure for the number of proteins that are translated in the cell; this is only approximately correct, but nevertheless an indication.

The recognition capability of a single-stranded RNA or DNA towards its complementary sequence can be exploited in a very simple way to measure the presence and amount of DNA in a given sample. In order to put this idea into practice, it has now become possible, using techniques from the microelectronics industry, to produce biochips - in analogy to microelectronic chips.

The principle of a *biochip*, or *DNA microarray* in our context, is the following. On a substrate, a collection of single-stranded DNA molecules, the *probes*, is built. (We will not bother here with the technical details how this is done; the interested reader is asked to consult the literature indicated at the end of the Chapter.) In the case of *Affymetrix chips*, the only type of microarrays we address here, the sequences contain 25 base pairs which have been selected from a gene of interest. In fact, about 10-16 probes are produced and fixed at the substrate, reflecting different selected subsequences from the same gene. In addition, for each perfectly matching probe - one that is exactly identical to a subsequence from the selected gene - a mismatching probe is produced. This probe differs from the perfect match by one nucleotide. The mismatch probe is a control, which is used to assess *cross-hybridizations*, i.e. non-specific hybridizations.

Based on this idea microarrays containing full genomes can now be built. They are used by exposing the probe strands to a set of target molecules which will bind to the selected probes, as schematically shown in Figure 2.12. The amount of bound DNA is measured, typically optically by using fluorescently labelled *target RNA*.

How do targets and probe meet each other on the chip? A very simple one is the following (E. CARLON and T. HEIM, 2005), motivated by previous work by G. A. HELD et al., 2003. After hybridizing targets and probes light intensities I of the fluorescent markers are measured. Distinguishing between specific, S, and non-specific hybridizations, N, we write

$$I(c, \Delta G) = S(c, \Delta G) + N + \epsilon \tag{2.33}$$

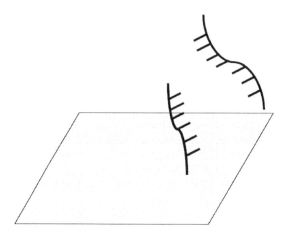

FIGURE II.2.12:   Hybridization on a microarray. Note that in the case of Affymetrix arrays, probes are DNA, while targets are RNA.

where ϵ stands for purely experimental noise. I is the intensity from the probe whose complementary target RNA is present in solution at a concentration c, and ΔG is the hybridization free energy - which we had discussed in Section 1.1. A distinction between mismatching or matching DNA does not need to be made, since both will differ only in their values of ΔG. The non-specific hybridization contribution N depends on total RNA concentration in solution, and probably other free energy parameters reflecting partial hybridizations. The value of N is not important if one decides to consider the quantity

$$\Delta I \equiv I(c) - I(0) \approx S(c, \Delta G) \tag{2.34}$$

Such a background subtraction is possible when we compare measurements in which c is the concentration of a particular gene; this is the case for the set of control measurements done in a *Latin square*, which consists of a well-defined dilution series for a set of specific genes added to a background, the so-called *'spike-in' genes*, and, importantly, contains the case $c = 0$.

   In the simplest case, which is well applicable to chains of the order of 25 base pairs as we know from Section 2.1 of this Chapter, we can understand the binding of the probe DNA and the target RNA as a two-state process, in which the target is either fully unbound in solution, or fully bound at the correct sequence at the surface. The binding is then described by a *Langmuir isotherm* of the form

$$S(c, \Delta G) = \frac{Ace^{-\beta\Delta G}}{1 + ce^{-\beta\Delta G}} \tag{2.35}$$

where, as before, $\beta = 1/RT$. The amplitude factor A sets an intensity scale;

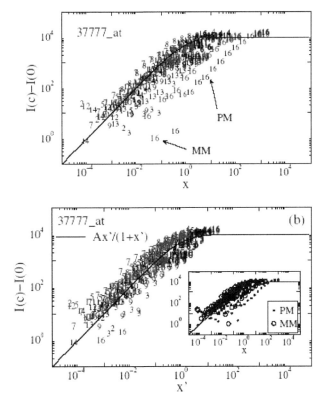

FIGURE II.2.13: Hybridization data falling on the Langmuir isotherm; plotted against the two scaling variables x (top) and x' (bottom) defined in the text. Reprinted with permission from E. CARLON and T. HEIM; Copyright (2005) by Elsevier.

it corresponds to the saturation value in the limit $c \gg e^{\beta \Delta G}$, i.e., whenever the concentration is high or the binding energy large.

Figure 2.13 shows an example of such data analysis. Noticing from eq.(2.35) that the free energy and the concentration appear always in the form of a product $x \equiv c \exp(-\beta \Delta G)$, we can use x as a scaling variable. The data then fall nicely on the Langmuir isotherms.

In these curves, the ΔG values were computed with parameters determined experimentally for DNA-RNA hybrids in solution and not for the DNA attached to the surface, and we may assume a systematic deviation between the

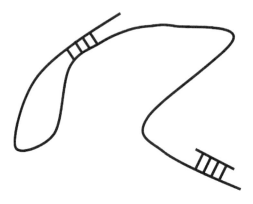

FIGURE II.2.14: Schematic drawing of RNA hybridization in solution indicating two possible effects, self-hybridization of targets and hybridization of different fragments.

(known) solution values and the (unknown) surface values.

This systematic deviation between the free energy values in solution and at the surface can likewise be shifted into one overall parameter: temperature. Temperature T then changes status from an experimental to a fitting parameter. If its value is chosen to lie at around 700 K, i.e., roughly twice as large as the true experimental temperature, a very good agreement between experiment and the simple theory is reached.

As seen in Figure 2.13, only the data from probe 16 deviate significantly from the computed Langmuir isotherm, but they seem to follow another isotherm which appears shifted to the right. This behaviour can be captured by introducing a probe-dependent parameter

$$\alpha_k = (1 + \tilde{c}\exp(-\beta\Delta G_{RNA}))^{-1}, \tag{2.36}$$

which takes into account the effect of RNA-RNA hybridization in solution, leading to secondary structures of the RNA strands closing in on themselves, thereby reducing the amount of available sequences for binding at the surface. This effect can be taken into account in the parameter combination $\tilde{c}\exp(-\beta\Delta G_{RNA})$, wherein $\tilde{c}$ is a fitting parameter. Figure 2.13 displays the full data collapse if the scaling variable

$$x' = \alpha_k c \exp{-\beta\Delta G} \tag{2.37}$$

is used. What happens during hybridization in solution is indicated schematically in Figure 2.14. Sequences with high free energies tend to hybridize to other chains present in the solution, and they are thus not available anymore for the hybridization process at the surface - their 'active' concentration is reduced as compared to that of other sequences.

Thus, at the end of the chapter on DNA, we see how relevant the formation of secondary structures can be even for DNA-RNA microarrays. The conclusion is that we need to learn more about this process, and we are thus ready to leave DNA for the RNA world.

Additional Notes

Short DNA's remain of interest due to their applicability in biotechnology, in particular in microarrays. Modifications of the linear DNA have become of interest as well; the simplest such systems are DNA hairpins, in which a single-stranded DNA bends back to itself and increases specificity, the so-called *molecular beacons* (G. BONNET et al., 1999). Short chains can also be used to decorate nanoparticles which enable to establish a link between molecular recognition and phase behaviour (D. B. LUKATSKY and D. FRENKEL, 2004).

Finite-size scaling. Real DNA sequences always have a finite length L, thus one may wonder whether the whole discussion of phase transition effects is at all applicable. There is a systematic way to extract the effect of finite system size at phase transitions which runs under the name *finite-size scaling*; for the Poland-Scheraga model, this has only recently been discussed (L. SCHÄFER, 2005). The conclusions are not much different from what was discussed here, though.

Other models used in studies of DNA denaturation. There are several other models that have been developed for the description of DNA denaturation aside from the Poland-Scheraga model. We here comment only on one model which follows a rather different philosophy: the Dauxois-Peyrard-Bishop model by TH. DAUXOIS et al., 1993, following earlier work by M. PEYRARD and A. R. BISHOP, 1989. Their model is more microscopic than the Poland-Scheraga model; it starts out from the dynamics of the chains, i.e., the main ingredient in the model is the transverse stretching y_n of the hydrogen bonds between the complementary bases counted by the index n. The model is defined by the Hamiltonian

$$H = \sum_n \left[\frac{1}{2} m \dot{y}_n^2 + V(y_n) \right]$$
(2.38)

where m is the mass of the bases in the kinetic energy. The potential model contains two contributions

$$V(y_n) = \left[D_n \left(e^{-\alpha_n y_n} - 1 \right)^2 + \frac{k}{2} \left(1 + \varrho e^{(y_n + y_{n-1})} \right) (y_n - y_{n-1})^2 \right].$$
(2.39)

The Morse potential in the first term describes the effective interactions between complementary bases: it contains both the attraction due to the hydrogen bonds forming the base pairs and the repulsion of the negatively charged phosphates in the backbone of the strands, which is screened by the surrounding solvent. The parameters D_n and α_n distinguish between the two complementary base pair combinations at site n, and hence induce a sequence dependence. The second term comprises the stacking interactions. The exponential term modifies an otherwise harmonic potential. This nonlinearity

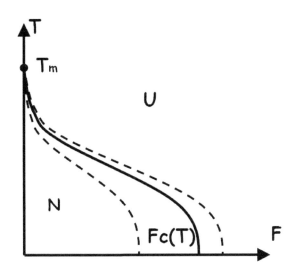

FIGURE II.2.15: Unzipping and thermal denaturation phase diagram

is essential: representing local constraints in nucleotide motions, it induces long-range cooperative effects. The stiffening of the coupling in the bound state compared to that in the open state leads to an abrupt entropy-driven transition. Similar conclusions were drawn by D. CULE and T. HWA, 1997.

The Dauxois-Peyrard-Bishop-model has, as the Poland-Scheraga model, also been confronted with experimental data, in particular for short sequences (A. CAMPA and A. GIANSANTI, 1998). Moreover, it has recently been used to correlate the dynamics of loop openings in the chain with regulatory sequence signatures. Correlations of regions with a high probability of loop openings with transcription start sites have been reported (C. H. CHOI et al. 2004 and G. KALOSAKAS et al., 2004). More recent results are more skeptical, see T. S. VAN ERP et al., 2005.

DNA unzipping. The opening of DNA does not necessarily have to arise by thermal denaturation - which typically occurs at non-physiological temperatures, and hence cannot play any direct role in a living cell. Other factors, such as a change of pH, or the direct application of a local force have, however, a similar effect on the molecule. As we have seen already in Part I, pulling experiments on DNA have become a standard approach due to the development of new methods of micromanipulation. From the point of view of the denaturation phase transition, the opening of the chains under the action of a force has been termed 'unzipping transition' (for obvious reasons). It can be shown to be a first-order transition, as much as the phase transition in the

Ising model in a magnetic field, see Part I, Chapter 1.

Theoretically, the unzipping transition can be described by a variant of the Worm-Like Chain model, in which in addition to the bending energy and the applied force a binding potential between the two chains is taken into account (for a brief review, see S. COCCO et al., 2002). From the calculation of the free energy of this model one finds a phase diagram in the temperature-force plane, shown in Figure 2.15, which qualitatively indicates the region in which the transitions occur. This figure is only of limited generality: the precise location of the first-order unzipping line can be anywhere inside the range set by the dashed lines, since it is in general sequence-dependent.

DNA microarrays. Despite the enormous interest in the use of DNA microarrays, many of their basic physical properties remain fairly poorly understood; an overview of hybridization protocols before the arrival of the high-throughput techniques is by J. G. WETMUR, 1991. The mechanism of probe-target encounter relies on the efficient diffusion of the target molecules on the surface; kinetic modelling efforts have been devoted to model the diffusion problem, see D. ERICKSON et al., 2003. It is of utmost importance that the fluorescence measurements are done when thermal equilibrium has been reached, as pointed out by G. BHANOT et al., 2003. Conceptually basic modelling approaches based on the Langmuir isotherm approach are reviewed by A. HALPERIN et al, 2006. The paper that inspired the approach discussed here is by G. A. HELD et al., 2003. It should be compared with papers from a more statistical approach, as e.g. L. ZHANG et al., 2003. Physics-based approaches tend to be based on few assumptions, and a small number of parameters, quite in contrast to purely statistical approaches.

References

G. Altan-Bonnet, A. Libchaber and O. Krichevsky, *Bubble Dynamics in Double-Stranded DNA*, Phys. Rev. Lett. **90**, 138101 (2003)

J. Bayer, J. O. Rädler and R. Blossey, *Chains, Dimers, and Sandwiches: Melting Behaviour of DNA Nanoassemblies*, Nano Letters **5**, 497-501 (2005)

G. Bhanot, Y. Louzoun, J. Zhu and C. DeLisi, *The Importance of Thermodynamic Equilibrium for High Throughput Gene Expression Arrays*, Biophys. J. **84**, 124-135 (2003)

R. D. Blake, *Cooperative Lengths of DNA During Melting*, Biopolymers **26**, 1063-1074 (1987)

R. D. Blake and S. G. Delcourt, *Thermal Stability of DNA*, Nucl. Acids Res. **26**, 3323-3332 (1998)

R. D. Blake, J. W. Bizarro, J. D. Blake, G. R. Day, S. G. Delcourt, J. Knowles, K. A. Marx and J. Santa Lucia, Jr, *Statistical mechanical simulation of polymeric DNA melting with MELTSIM*, Bioinformatics **15**, 370-375 (1999)

R. Blossey and E. Carlon, *Reparametrizing the loop entropy weights: Effect on DNA melting curves*, Phys. Rev. E **68**, 061911 (2003)

G. Bonnet, S. Tyagi, A. Libchaber and F. R. Kramer, *Thermodynamic basis of the enhanced specificity of structured DNA probes*, Proc. Natl. Acad. Sci. USA **96**, 6171-6176 (1999)

A. Campa and A. Giansanti, *Experimental tests of the Peyrard-Bishop model applied to the melting of very short DNA chains*, Phys. Rev. E **58**, 3585-3588 (1998)

E. Carlon, E. Orlandini and A. L. Stella, *Role of Stiffness and Excluded Volume in DNA Denaturation*, Phys. Rev. Lett. **88**, 198101 (2002)

E. Carlon, M. L. Malki and R. Blossey, *Exons, introns and DNA thermodynamics*, Phys. Rev. Lett. **94**, 178101 (2005)

E. Carlon and T. Heim, *Thermodynamics of RNA/DNA hybridization in high density oligonucleotide microarrays*, Physica A **362**, 433-449 (2006)

C. H. Choi, G. Kalosakas, K. O. Rasmussen, M. Hiromura, A. R. Bishop and A. Usheva, *DNA dynamically directs its own transcription initiation*, Nucl. Acids Res. **32**, 1584-1590 (2004)

S. Cocco, J. F. Marko and R. Monasson, *Theoretical models for single-molecule DNA and RNA experiments: from elasticity to unzipping*, C. R. Physique **3**, 569-584 (2002)

D. Cule and T. Hwa, *Denaturation of Heterogeneous DNA*, Phys. Rev. Lett. **79**, 2375-2378 (1997)

T. Dauxois, M. Peyrard and A. R. Bishop, *Entropy-driven DNA denaturation*, Phys. Rev. E **47**, R44-R47 (1993)

D. Erickson, D. Li, and U. J. Krull, *Modeling of DNA hybridization kinetics for spatially resolved biochips*, Anal. Biochem. **317**, 186-200 (2003)

T. S. van Erp, S. Cuesta-Lopez, J.-G. Hagmann and M. Peyrard, *Can One Predict DNA Transcription Start Sites by Studying Bubbles?*, Phys. Rev. Lett. **95**, 218104 (2005)

M. Fixman and J. J. Freire, *Theory of DNA Melting Curves*, Biopolymers **16**, 2693-2704 (1977)

Th. Garel and H. Orland, *Generalized Poland-Scheraga model for DNA hybridization*, Biopolymers **75**, 453-467 (2005)

W. Gilbert, *The Exon Theory of Genes*, vol. LII Cold Spring Harbour Symposia on Quantitative Biology (1987)

A. Halperin, A. Buhot and E. B. Zhulina, *On the hybridization isotherms of DNA microarrays: the Langmuir model and its extension*, J. Phys.: Condens. Matter, to appear (2006)

A. Hanke and R. Metzler, *Comment on: "Why is the DNA denaturation transition first order?"*, Phys. Rev. Lett. **90**, 159801 (2003)

G. A. Held, G. Grinstein and Y. Tu, *Modeling of DNA microarray data by using physical properties of hybridization*, Proc. Natl. Acad. Sci. USA **24**, 7575-7580 (2003)

Y. Kafri, D. Mukamel and L. Peliti, *Melting and unzipping of DNA*, Eur. Phys. J. B **27**, 135-146 (2002)

G. Kalosakas, K.O. Rasmussen, A. R. Bishop, C. H. Choi and A. Usheva, *Sequence-specific thermal fluctuations identify start sites for DNA transcrip-*

tion, Europhys. Lett. **68**, 127-133 (2004)

D. B. Lukatsky and D. Frenkel, *Phase Behavior and Selectivity of DNA-Linked Nanoparticle Assemblies*, Phys. Rev. Lett. **92**, 068302 (2004)

R. Owczarzy, P. M. Vallone, F. J. Gallo, T. M. Paner, M. J. Lane, A. S. Benight, *Predicting Sequence-Dependent Melting Stability of Short Duplex DNA Oligomers*, Biopolymers **44**, 217-239 (1998)

M. Peyrard and A. R. Bishop, *Statistical Mechanics of a Nonlinear Model for DNA Denaturation*, Phys. Rev. Lett. **62**, 2755-2759 (1989)

D. Poland, *Recursion Relation Generation of Probability Profiles for Specific-Sequence Macromolecules with Long-Range Correlations*, Biopolymers **13**, 1859-1871 (1974)

D. Poland and H. A. Scheraga, *Theory of helix-coil transitions in biopolymers*, Academic Press (1970)

J. Santa-Lucia Jr., *A unified view of polymer dumbbell, and oligonucleotide DNA nearest-neighbor thermodynamics*, Proc. Natl. Acad. Sci. USA **95**, 1460-1465 (1998)

L. Schäfer, *Can Finite Size Effects in the Poland-Scheraga Model Explain Simulations of a Simple Model for DNA Denaturation*, preprint cond-mat/0502668 (2005)

R. Thomas, *Recherches sur la dénaturation des acides desoxyribonucléiques*, Biochim. Biophys. Acta **14**, 231-240 (1954)

R. M. Wartell and A. S. Benight, *Thermal denaturation of DNA molecules: a comparison of theory with experiment*, Phys. Rep. **126**, 67-107 (1985)

J. G. Wetmur, *DNA Probes: Applications of the Principles of Nucleic Acid Hybridization*, Crit. Rev. Biochem. Mol. Biol. **26**, 227-259 (1991)

E. Yeramian, *Genes and the physics of the DNA double-helix*, Gene **255**, 139-150 (2000)

E. Yeramian, *The physics of DNA and the annotation of Plasmodium falciparum*, Gene **255**, 151-168 (2000)

E. Yeramian, S. Bonnefoy and G. Langsley, *Physics-based gene identification: proof of concept for Plasmodium falciparum*, Bioinformatics **18**, 190-193

(2002)

L. Zhang, M. F. Miles and K. D. Aldape, *A model of molecular interactions on short oligonucleotide microarrays*, Nat. Biotech. **21**, 821-828 (2003); *Erratum* ibid, **21**, 841 (2003).

Chapter 3

RNA

3.1 Computing RNA secondary structure: combinatorics

We have seen in the previous chapters why the knowledge of RNA secondary structure is important - for information processing in biology and biotechnology. In this Chapter we want to learn how to predict RNA structure. Predicting RNA structure, as we will see, has three aspects. The first aspect is combinatorial: how can we classify and compute the *possible* structures? The second aspects relates to the question which of these structures are selected based on their free energy: from the set of all possible structures, which is the one that is energetically the most favorable one? The third and final aspect touches upon the kinetics of the fold formation: how does the molecule actually acquire its fold?

Maximal base pairing. We begin the discussion by formalizing the graphical representations introduced in Chapter 1.

If the primary structure of an RNA molecule - i.e., its sequence - of length n is denoted by a string

$$r = r_1...r_n , \qquad (3.1)$$

its secondary structure can be described as a set S of disjoint pairs (r_i, r_j) for $1 \leq i < j \leq n$. Considering the bases as vertices on a graph, and all pairings as the edges of that graph, the secondary structure is a matching in a graph $G = (V, E)$ with the properties that V contains a vertex for every base pair r_i, $i = 1, ..., n$, and E contains an edge (u, v) if and only if $u, v \in V$ are complementary bases.

In the prediction of secondary structure, we want to first exclude *pseudoknots*. In the above notation such a knot occurs when a base r_i is paired with a base r_j, and a base r_k with r_l such that $i < k < j < l$: the pairs overlap. It is this condition which we reject for the moment.

The prescription we have given allows to cover all possible secondary structures, except for the occurrence of pseudoknots. However, we would not know

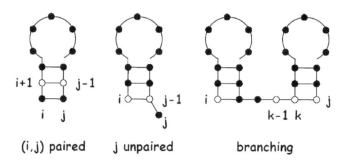

FIGURE II.3.1: Dynamic programming recursion for RNA structure (see text).

which one to select from the set of structures. A natural selection principle is based on free energy. Assuming that we can associate with every complementary base pair an energy contribution $a(r_i, r_j) = -1$ and put $a = 0$ otherwise, we see that the minimal (free) energy $E(S) = \sum_{(r_i, r_j)} a(r_i, r_j)$ will be attained for the RNA structure with the maximal number of base pairs.

In this simplified version which treats all bound pairs on equal footing, the problem of RNA secondary structure determination can be solved by a dynamic programming approach. With the assumption that $a(r_i, r_j)$ is independent of all other pairs and positions of r_i and r_j in the sequence, we can compute the free energy of a substring $r_i...r_j$ of the RNA molecule disregarding the surrounding $r_1...r_{i-1}$ and $r_j + 1...r_n$. In this way, we can use solutions for smaller strings to obtain those for larger strings and obtain a recursive solution.

The resulting dynamic programming algorithm is easy to develop; the possibilities that can arise in building up the structure are shown in Figure 3.1. There are three alternatives:

- Add paired bases to an optimal structure for the subsequence $i+1, j-1$;

- Add an unpaired base to an optimal structure for the subsequence $i, j-1$; this possibility arises symmetrically for the other end of the molecule;

- Combine two optimal substructures $i, k-1$ and k, j.

The latter step obviously runs into problems when pseudoknots are to be accounted for.

The resulting recursion can be summarized as

$$E(S_{i,j}) = \min \begin{cases} E(S_{i+1,j-1}) - 1 \\[2mm] E(S_{i+1,j}) \\[2mm] E(S_{i,j-1}) \\[2mm] \min\{E(S_{i,k-1}) + E(S_{k,j})\}, \quad i < k \le j \end{cases} \qquad (3.2)$$

In order to initialize the algorithm, one requires $E(S_{i,i}) = 0$ for $i = 1, .., n$ and $E(S_{i,i-1}) = 0$ for $i = 2, .., n$.

Let us illustrate this scheme with a small example (S. R. EDDY, 2004). We take the RNA sequence $r = GGGAAAUCC$; for each allowed pair AU, GC we take $\alpha = -1$, and $\alpha = 0$ otherwise. The pairing matrix $(r \times r)$ at the initialisation step reads

$$\begin{bmatrix} & G\,G\,G\,A\,A\,A\,U\,C\,C \\ G & 0 \\ G & 0 \;\; 0 \\ G & x \;\; 0 \;\; 0 \\ A & x \;\; x \;\; 0 \;\; 0 \\ A & x \;\; x \;\; x \;\; 0 \;\; 0 \\ A & x \;\; x \;\; x \;\; x \;\; 0 \;\; 0 \\ U & x \;\; x \;\; x \;\; x \;\; x \;\; 0 \;\; 0 \\ C & x \;\; x \;\; x \;\; x \;\; x \;\; x \;\; 0 \;\; 0 \\ C & x \;\; x \;\; x \;\; x \;\; x \;\; x \;\; x \;\; 0 \;\; 0 \end{bmatrix} \qquad (3.3)$$

where the lower half (x) is not used, and the diagonal and its left neighbour are put to 0.

In the computation, one finds that entry Y depends on all entries denoted by y, and $w = (i + 1, j)$,

$$\begin{bmatrix} & G\,G\,G\,A\,A\,A\,U\,C\,C \\ G & 0 \\ G & 0 \;\; 0 \\ G & x \;\; 0 \;\; 0 \;\; y \;\; y \;\; y \;\; y \;\; Y \\ A & x \;\; x \;\; 0 \;\; 0 \qquad\quad w \;\; y \\ A & x \;\; x \;\; x \;\; 0 \;\; 0 \qquad\quad y \\ A & x \;\; x \;\; x \;\; x \;\; 0 \;\; 0 \qquad y \\ U & x \;\; x \;\; x \;\; x \;\; x \;\; 0 \;\; 0 \;\; y \\ C & x \;\; x \;\; x \;\; x \;\; x \;\; x \;\; 0 \;\; 0 \\ C & x \;\; x \;\; x \;\; x \;\; x \;\; x \;\; x \;\; 0 \;\; 0 \end{bmatrix} \qquad (3.4)$$

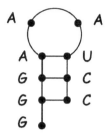

FIGURE II.3.2:   Exemplary RNA structure obtained from the simple algorithm.

so that one obtains the following result by going through the recursion

$$
\begin{bmatrix}
 & G & G & G & A & A & A & U & C & C \\
G & 0 & 0 & 0 & 0 & 0 & 0 & -1 & -2 & (-3) \\
G & 0 & 0 & 0 & 0 & 0 & 0 & -1 & -2 & (-3) \\
G & x & 0 & 0 & 0 & 0 & 0 & -1 & (-2) & -2 \\
A & x & x & 0 & 0 & 0 & 0 & (-1) & -1 & -1 \\
A & x & x & x & 0 & 0 & (0) & -1 & -1 & -1 \\
A & x & x & x & x & 0 & (0) & -1 & -1 & -1 \\
U & x & x & x & x & x & 0 & 0 & 0 & 0 \\
C & x & x & x & x & x & x & 0 & 0 & 0 \\
C & x & x & x & x & x & x & x & 0 & 0
\end{bmatrix}
\qquad (3.5)
$$

The minimum structure can be found by tracing back through the table following a diagonal trace (i, j) to $(i + 1, j - 1)$. The result is shown in Figure 3.2; the trace back is indicated by terms in brackets. The complexity of the algorithm to compute the structure is $O(n^3)$, since there are n^2 entries, and each computation is $O(n)$. The trace back to find the structure takes again linear time $O(n)$, if back pointers are used.

Following the original approach by R. NUSSINOV and A. B. JACOBSON, 1980, there have been several further developments of algorithms for RNA secondary structure prediction. M. ZUKER, 1989, made an important contribution by developing a tool for the determination of sub-optimal folds. More recent developments are targeted towards the inclusion of pseudoknots; we comment on them in the Additional Notes.

The basic approach to RNA secondary structure, as we have seen, is to compute the possible combinations by a recursion, and to score them by a suitably defined free energy. Thinking back to what we saw for DNA, one might question whether the first part - the computation of the configurations - cannot also be done based on our main tool from statistical mechanics, the partition function Z. This is what follows.

3.2   The RNA partition function

The partition function of a chain of L nucleotides can be written in the following very general way as (H. ORLAND and A. ZEE, 2002)

$$\mathcal{Z} = \int \prod_{k=1}^{L} d^3 \mathbf{r}_k \, \mathcal{F}(\{\mathbf{r}\}) \, Z_L(\{\mathbf{r}\}) \,. \tag{3.6}$$

Here, $\mathbf{r}_k$ is the position vector of the k-th nucleotide. The function

$$\mathcal{F}(\{\mathbf{r}\}) = \prod_{i=1}^{L-1} f(|\mathbf{r}_{i+1} - \mathbf{r}_i|) \tag{3.7}$$

is a model-dependent function of the molecular geometry, and takes into account the steric constraints of the chain. Standard choices for f are

$$f(r) = \delta(r - \ell) \,, \tag{3.8}$$

if the nucleotides are connected by rigid rods of size ℓ, or

$$f(r) = \exp[-(r - \ell)^2/6\eta^2] \tag{3.9}$$

if the rods are taken as elastic springs with η as a measure of their stiffness.

The partial partition function Z_L in eq.(3.6) counts the different configurations of paired bases. It can be defined by the series

$$Z_L(\{\mathbf{r}\}) = 1 + \sum_{(ij)} V_{ij}(\mathbf{r}_{ij}) + \sum_{(ijkl)} (V_{ij}(\mathbf{r}_{ij})V_{kl}(\mathbf{r}_{kl}) + V_{ik}(\mathbf{r}_{ik})V_{jl}(\mathbf{r}_{jl})) + \dots$$
$$\tag{3.10}$$

with $\mathbf{r}_{ij} \equiv |\mathbf{r}_i - \mathbf{r}_j|$, and where the summation index (ij) denotes all pairs with $i < j$, $(ijkl)$ all quadruplets with $i < j < k < l$, and so forth. The first term of the series describes the binding energy between the bonds i and j, with the indices running all along the chain, the second the binding of i and j in a configuration together with k and l, and so forth.

The factors

$$V_{ij} \equiv \exp(-\beta \varepsilon_{ij} v_{ij}(\mathbf{r}_{ij}))\theta(|i - j| > 4) \tag{3.11}$$

are the Boltzmann factors associated with a (4×4)-dimensional symmetric matrix ε_{ij} of bond energies between the i-th and j-th bases at a distance $\mathbf{r}_{ij}$; $\beta = 1/(k_B T)$ is the inverse thermal energy as before. In eq.(3.11), the factor $v_{ij} = v(|\mathbf{r}_i - \mathbf{r}_j|)$ is a short-range attractive interaction between the bases. The Heaviside function $\theta(|i - j| > 4)$ expresses the steric constraint which prohibits hybridization of the bases in closest proximity to each other. Finally, note that $V_{ii} = 0$.

The series (3.10) can be expressed in integral form by introducing a set of $i = 1, ..., L$ *Hermitian matrices*[1] φ_i, $i = 1, ..., L$, using the expression for the ordered matrix product

$$\mathcal{O}_\pi[\{\varphi\}] \equiv \prod_{l=1}^{L}(1 + \varphi_l) \equiv (1 + \varphi_1)(1 + \varphi_2) \cdots (1 + \varphi_L). \qquad (3.12)$$

The matrices have the dimension $(N \times N)$. The parameter N is introduced on purely formal grounds; as will be seen below, it allows to organize the expansion of the partition function according to the *topology of RNA*. It can, however, also be related to physical quantity, the concentration of a chemical which favors pseudoknots. Such chemicals, e.g., are divalent ions such as Mg^{2+}, and the relationship is given via the idenfication $N^{-2} \equiv \exp \beta\mu$ where μ is the chemical potential of the ions.

Coming back to the calculation, the resulting formula for $Z_L(N)$ is

$$Z_L(N) = A_L(N)^{-1} \int \prod_{k=1}^{L} d\varphi_k e^{-\frac{N}{2}\sum_{ij}(V^{-1})_{ij}Tr(\varphi_i\varphi_j)} \frac{1}{N} Tr\mathcal{O}_\pi[\{\varphi\}] \quad (3.13)$$

where the normalization factor is given by

$$A_L(N) = \int \prod_{k=1}^{L} d\varphi_k e^{-\frac{N}{2}\sum_{ij}(V^{-1})_{ij}Tr(\varphi_i\varphi_j)}. \qquad (3.14)$$

In both expressions, V is an $(L \times L)$-matrix with entries V_{ij}.

After the introduction of the matrices, the partition function given by e-q. (3.13) looks like a Gaussian integral in the matrices φ_k over a product observable in the φ_k. The product of the terms $(1 + \varphi_l)$ evaluates into a polynomial of order L and hence we have to perform Gaussian integrals over all the contributions of this polynomial from order 1 to L - thus nothing but

---

[1]A square matrix is hermitian if it is *self-adjoint*, i.e., fulfills $A_{ij} = \overline{A}_{ji}$, where the overbar stands for the complex conjugate. We encountered examples of self-adjoint matrices in Part I, Chapter 1, in the discussion of the Ising model, e.g., the matrix σ_3.

the moments of the partition function.

The important thing is now that the introduction of the matrix-dimension N gives a handle to reorganize the series representation. One can show that the result is an asymptotic series in $1/N$ of the form

$$Z_L(N) = 1 + \sum_{i<j} V_{ij} + \sum_{i<j<k<l} V_{ij}V_{kl} + \dots + \frac{1}{N^2}\sum_{i<j<k<l} V_{ik}V_{jl} + \dots \quad (3.15)$$

The comparison of the two expressions eqs.(3.10), (3.15) shows that both coincide for $N = 1$; the latter, however, for $N > 1$ now contains information about the *topology of RNA*. The $O(1)$-terms of eq.(3.15) yield the planar secondary structures of RNA, while the $1/N^2$-terms correspond to RNA tertiary structure.

The secondary structure can be evaluated within this formalism by approximating the integral (3.13) by its saddle-point value in the limit $N \to \infty$ (see Part I, Chapter 1). In order to be able to perform this limit, the integral has to be transformed by a Hubbard-Stratonovich transform (again, see Part I, Chapter 1) into an expression in which the dependence on the parameter N - which is still the dimension of the matrices - becomes explicit. How such a transform is set up can be found in Part I, Chapter 1; it is left as a *Task* for the readers to apply it to the present case. Those who want to see it explicitly can find it in the paper by (H. ORLAND and A. ZEE, 2002), which also discusses how eq.(3.15) is obtained.

The result of the calculation is the expression (with C as an irrelevant normalization factor)

$$Z_L(N) = \frac{1}{C}\int dA e^{-\frac{N}{2}Tr\, A^2 + NTr\, \ln M(A)} M^{-1}(A)_{L+1,1} \quad (3.16)$$

where the integral runs over all Hermitian matrices A of dimension $(L+1) \times (L+1)$. M is a matrix function of A given by

$$M_{ij} = \delta_{ij} - \delta_{i,j+1} + i\sqrt{V_{i-1,j}}A_{i-1,j} \quad (3.17)$$

and the symbol for trace, Tr, means - as in Part I, Chapter 1 - the sum over the diagonal elements.

The saddle-point of eq.(3.16) follows from the variation $\delta S(A)/\delta A = 0$, where S is

$$S(A) \equiv \frac{1}{2}Tr\, A^2 - Tr\, \ln M(A). \quad (3.18)$$

The stationary point is given by

$$A_{lk}^0 = i\sqrt{V_{lk}}(M^{-1})_{l,k+1} \,. \tag{3.19}$$

Introducing $G_{ij} \equiv (M^{-1})_{i+1,j}$, and using the identity $\sum_j M_{ij}(M^{-1})_{jk} = \delta_{ik}$ one obtains the so-called *Hartree equation*

$$G_{i+1,k} = \delta_{i+2,k} + G_{ik} + \sum_j V_{i+1,j}G_{i,j+1}G_{j-1,k} \,. \tag{3.20}$$

The Hartree equation is a recursion relation which can be solved for the boundary condition $G_{i,i+l} = 0$ for $l \geq 0$. Then, G_{ij} is the partition function of the secondary structure of a chain starting at base j and ending at base i, just as it is used in dynamic programming algorithms. We have therefore, as the saddle-point of the partition function, recovered a formal expression which allows to compute RNA secondary structure. We can use the result eq.(3.20) to compute, recursively, the structure of RNA; note that the energy evaluation is contained in V.

From this result, we can now branch off into two directions: we apply eq. (3.20), or we go on to evaluate the next terms in the series. This would amount to go for pseudoknots. We will do both, and begin with the pseudoknots.

Pseudoknots. The computation of pseudoknots based on the theory described in the previous section is rather involved, since the computation of the higher order terms in $1/N^2$ requires the use of field-theoretic methods which are out of the scope of this book.[2] The interested reader is asked to consult the original literature at this point.

Here, in what follows, we want restrict ourselves to obtain a general idea of the occurrence of pseudoknots, based on the topological theory. The basic question we address is: as the RNA structures become more and more complex, how many pseudoknots become will arise? This theory allows us to readily determine the *number of RNA pseudoknots* according to their topological character, and to see how it evolves as the sequence increases in length. For this calculation, some simplifying assumptions can be made. If any possible pairing between nucleotides is allowed (independent of the identity of the base pair and its location along the chain), and if all of these pairings will occur with equal probability, then the matrix V_{ij} has identical entries, which we suppose as $v > 0$ for all i, j.[3]

---

[2] We were pushing the limits a bit already. The method used here is well-known as a $1/N$-expansion in the context of field theory in statistical mechanics and the theory of elementary particles, and is certainly not easy to digest for a bioinformatician.

[3] A technical detail: in order for the computations to make sense when we do this, we need to make V_{ij} positive definite. We can do this by adding an arbitrary real number a to

The computation of the integral $Z_L(N)$ in this simplified case runs as follows. The original expression for $Z_L(N)$, eq.(3.15), can be rewritten using the integral transforms introduced before as

$$Z_L(N) = A^{-1}(N) \int d\sigma e^{-\frac{N}{2v}Tr\sigma^2} \frac{1}{N} Tr(1+\sigma)^L , \qquad (3.21)$$

where σ is a single $(N \times N)$-matrix. The normalization factor in this case is explicitly given by

$$A(N) = \int d\sigma e^{-\frac{N}{2v}Tr\sigma^2} = \left(\frac{\pi v}{N}\right)^{\frac{N^2}{2}} 2^{N/2} . \qquad (3.22)$$

In order to solve the Gaussian matrix integral (3.21) it is convenient to introduce the *spectral density* ϱ_N of the matrix σ,

$$\varrho_N(\lambda) \equiv A^{-1}(N) \int d\sigma e^{-\frac{N}{2v}Tr\sigma^2} \frac{1}{N} Tr\delta(\lambda - \sigma) . \qquad (3.23)$$

This is a convenient trick - in the same spirit as the Hubbard-Stratonovich transformation we used for the Ising model in Part I, Chapter 1. With the spectral density we can represent $Z_L(N)$ as

$$Z_L(N) = \int_{-\infty}^{+\infty} d\lambda \varrho_N(\lambda)(1+\lambda)^N , \qquad (3.24)$$

where the identity $\int_{-\infty}^{+\infty} d\lambda\, \varrho_N(\lambda) = 1$ was used in eq.(3.21). Hence, we have reduced a multidimensional problem to a one-dimensional integral.

One can now introduce the (exponential) generating function of $Z_L(N)$,

$$G(t, N) \equiv \sum_{L=0}^{\infty} Z_L(N)\frac{t^L}{L!} = \int_{-\infty}^{+\infty} d\lambda \varrho_N(\lambda) e^{t(1+\lambda)} , \qquad (3.25)$$

and we need the explicit form of $\varrho_N(\lambda)$. The latter is known from *Random Matrix Theory* (see M. L. MEHTA, 1991). It can be expressed in terms of a series of Hermite polynomials $H_k = (-1)^k \exp(x^2)(d^k/dx^k)\exp(-x^2)$,

$$\varrho_N(\lambda) = \frac{e^{-N\lambda^2/(2v)}}{\sqrt{2\pi v N}} \sum_{k=0}^{N-1} \binom{N}{k+1} \frac{H_{2k}(\lambda\sqrt{N/2v})}{2^k k!} . \qquad (3.26)$$

With this one obtains for $G(t, N)$ the expression

$$G(t, N) = e^{\frac{vt^2}{2N} + t} \frac{1}{N} L_{N-1}^{(1)}\left(-\frac{vt^2}{N}\right) , \qquad (3.27)$$

the diagonal elements. Since no diagonal terms appear in the original series ($V_{ii} = 0$), this number plays only a formal regularizing role, and the final result can be shown to be independent of its choice.

where $L_N^{(1)}(z)$ are the generalized *Laguerre polynomials*.[4] The series expansion of $G(N, t)$ in t now gives the first coefficients of $Z_L(N)$. Putting $v = 1$, one can write

$$Z_L(N) = \sum_{L=0}^{\infty} \frac{a_{L,g}}{N^{2g}} \tag{3.28}$$

where the coefficients $a_{L,g}$ determine the number of diagrams at fixed length L and *fixed genus g*.

Now that we have mastered this excursion into the functions of classical mathematical physics, we want to see what we can learn from this result. How can we interpret it?

Let's look at an example. For $L = 8$, we have

$$Z_8(N) = 1 + 28v + 140v^2 + 140v^3 + 14v^4 + \frac{1}{N^2}(70v^2 + 280v^3 + 70v^4) + 21\frac{v^4}{N^4} \tag{3.29}$$

The meaning of the numbers is:

- The power of v is the number of nucleotides that are paired;

- The power of N^{-2} is the genus g of the structure (i.e., the genus of the surface on which it can be drawn without bond crossings);

- Putting $v = 1$, one obtains the number of structures for a given genus g;

- Putting $N = 1$, one obtains the number of structures with a given number of bonds.

Thus, there are 28 planar structures with one bonded pair, 70 structures with two bonded pairs on a torus, etc. The number of structures with genus $g = 1$ is 420, while the number of structures with four pairs (the maximal number) is 105.

After this example, we finish by reading off some general characteristics from formula (3.28). An analysis of this series for a given length $L \gg 1$ shows

---

[4]The generalized Laguerre polynomials are defined by

$$L_n^{\alpha}(x) = \frac{(\alpha + 1)_n}{n!}[_1F_1(-n; \alpha + 1; x)]$$

where $(\alpha)_n = \Gamma(x + n)/\Gamma(n)$ is the *Pochhammer symbol* and $[_1F_1(a; b; z)] = \sum_{k=0}^{\infty} \frac{(a)_k}{(b)_k} \frac{z^k}{k!}$ is the *confluent hypergeometric function of the first kind*.

that the normalized number of pseudoknots $a_{L,g}/Z_L(1)$ is always peaked at
a characteristic genus of roughly $g_c(L) \sim 0.23L$. For fixed L, the maximally
achievable genus comes out to lie at $g \leq L/4$. More interestingly, in order for
a structure to have a given genus g, it needs to have a length $L \geq 4g$. I.e., for
an RNA structure with $g = 7$, the sequence must be at least 28 bases long.

Task. Do a literature search and find out what is the RNA structure with
the largest number of pseudoknots, hence highest genus currently known.

3.3   RNA phase behaviour and folding kinetics

The RNA molten and glass phases. In the previous sections we have
seen that one can consider the RNA folding problem as one of combinatorics
and energetics: the correct fold of an RNA molecule is the one which min-
imizes the free energy. We have not, as we did before for DNA, put in the
specific base pairing energies, but this can easily be done, starting from the
Hartree equation (3.20).

However, even if we do this, a more fundamental problem arises. For DNA,
we have looked at the pairing of two complementary strands, which always
gives a well-defined minimum free energy configuration: there was not even
a question as to what the structure is. But for RNA that is not clear. If the
strands become longer and longer, the number of possible structures explodes.
And many of these configurations can be at least almost degenerate: there
will be several configurations with the same or nearly te same free energy.
How accessible - and how relevant - is thus the minimum energy fold? Maybe
we need to know also those structures that are not minimal in free energy,
but close? Isn't it possible that the minimum configuration is well-separated
from a nearby configuration by an energy barrier, and the molecular configuration, once trapped in the other minimum, will never change into the
true minimum? Such a high barrier would exist, e.g., if the two energetically competing structures would be structurally far apart in configuration space.

In order to address such kind of questions we have to discuss the *free energy
landscape* of the RNA structures.

The free-energy landscape of RNA turns out to be strongly temperature-
dependent. For high enough temperatures - but still below the denaturation
temperature of a folded RNA, since we do not want to break the structure -,
one may be allowed to ignore sequence-dependence to a first approximation,

as we did before.[5] We want to do this for the planar (non-pseudoknotted) structures and go back to the saddle-point approximation.

Under the assumption that sequence-dependence is negligible, the Hartree equation (3.20) becomes translationally invariant along the chain of length L and takes on the simplified form

$$G(L+1) = G(L) + q \sum_{k=1}^{L-1} G(k)G(L-k) \tag{3.30}$$

with $q \equiv exp(-\beta\epsilon_0)$. By using the transform

$$\widehat{G}(z) = \sum_{L=1}^{\infty} G(L)z^{-L} \tag{3.31}$$

the convoluted sum in eq.(3.30) can be eliminated, which gives rise to an algebraic equation for $\widehat{G}(z)$,

$$z\widehat{G}(z) - 1 = \widehat{G}(z) + q\widehat{G}^2(z). \tag{3.32}$$

This equation has the solution

$$\widehat{G}(z) = \frac{z - 1 - \sqrt{(z-1)^2 - 4q}}{2q}. \tag{3.33}$$

In order to transform back to $G(L)$, again a saddle-point approximation can be used. We then find the expression

$$G_0(L) \approx g_0(q)L^{-3/2}exp(-Lf_0(q)) \tag{3.34}$$

with $f_0(q) = -\ln(1 + 2\sqrt{q})$.

The result (3.34) shows the coexistence of an exponentially large number of RNA secondary structures with *equal* free energy. This phase is dominated by the configurational entropy of the molecules; it has been termed the *'molten' phase* (R. BUNDSCHUH and T. HWA, 2002).

How physically realistic is this phase? In order to check this, one has to vary the binding energies. In doing so we introduce what can be called *sequence disorder*, since we start from a molecule with identical bindings. We choose binding energies ε_{ij} with

$$\varepsilon_{i,j} = \begin{cases} -u_m\,(i,j)\ , & WC \\ u_{mm}\quad,\quad else \end{cases} \tag{3.35}$$

---

[5]But then based on a different argument since wanted to look at the different possible topologies.

with $u_m, u_{mm} > 0$, and WC stands for a Watson-Crick base. The $\varepsilon_{i,j}$ are taken as independent Gaussian distributed random variables with mean $\bar{\varepsilon}$ and variance

$$\overline{(\varepsilon_{i,j} - \bar{\varepsilon})(\varepsilon_{k,l} - \bar{\varepsilon})} = D\delta_{i,k}\delta_{j,l}. \tag{3.36}$$

For this choice it turns out that the molten phase is indeed thermodynamically stable for weak sequence disorder (R. BUNDSCHUH and T. HWA, 2002), but if we allow for arbitrarily strong sequence disorder the molten phase becomes unstable with respect to another phase: the *glass phase*. This phase is dominated by one or few structures of lower free energies.

How can one see that the molten phase must be unstable with respect to the glass phase? We follow an argument developed by R. BUNDSCHUH and T. HWA, 2002.

The argument is based on the introduction of a so-called *division free energy* in the molten phase, given by

$$\Delta F \equiv -k_B T \ln\left[\frac{G_0^2(L/2)}{G_0(L)}\right] \approx \frac{3}{2}k_B T \ln L. \tag{3.37}$$

It is the free energy cost of cutting a chain of length L into two non-interacting halves. This quantity characterizes the loss of configurational entropy a secondary structure of the molten phase will undergo by cutting.

Let us now consider the division free energy for an arbitrary base sequence. For each sequence, we select a segment $\ell \ll L$ of Watson-Crick-paired bases r_i such that the sequence $r_i...r_{i+\ell-1}$ is in the first half of the molecule, while $r_{j-\ell+1}...r_j$ is in the second half. For a random sequence of length L one can show rigorously that $\ell = \ln L/\ln 2$. We further observe that

$$\Delta F \equiv F_{div} - F_{free} \geq F_{div} - F_{paired} \tag{3.38}$$

where F_{paired} is the free energy of the ensemble of structures in which the complementary segments are paired and the remaining substrands,

$$F_{paired} = -\ell u_m + (N - 2\ell)f_0 + \frac{3}{2}k_B T[\ln L_1 + \ln L_2] \tag{3.39}$$

where L_1, L_2 are the lengths of the two remaining substrands. Taking as F_{div} the value of the molten phase,

$$F_{div} = f_0 L + 2(3/2)k_B T \ln L \tag{3.40}$$

(ignoring non-L dependent terms), and assuming that L_1, L_2 scale linearly with L, one is left with the estimate

$$\frac{3}{2}k_B T \geq \frac{u_m + 2f_0}{\ln 2}. \tag{3.41}$$

To close the argument we have to discuss the *low-temperature limit* of eq.(3.41). For $T \rightarrow 0$, $f_0 \rightarrow -\bar{n}u_m$, where $\bar{n}$ is the average number of base pairs per monomer in the minimal free energy structure; if all bases are paired, we have $\bar{n} = 1/2$. Since there will in general be a finite fraction of non-paired bases, we expect $\bar{n} < 1/2$. Thus, by increasing the value of u_m while decreasing β, the inequality (3.41) must be violated at a critical value u_m^*, necessitating the existence of a phase of secondary structures at a lower temperature. This then is the 'glass phase'.

For real RNA sequences, apart from the glass phase, we also have to consider the 'native phase', which corresponds to the minimum free energy solution. One can characterize the structural phases of RNA are characterised by the average sizes of the molecules based on the scaling behaviour of the average diameter $\langle h \rangle$. Supposing a relation

$$\langle h \rangle \sim L^m \tag{3.42}$$

between diameter $\langle h \rangle$ and sequence length L, introducing an exponent m, one finds that the exponent varies between $0.5 < m < 1$ in the glass, molten and native structures. Thus, $\langle h \rangle$ displays only a rather weak dependence on L.

RNA folding kinetics. Given that the configuration space of RNA is characterized by a rugged landscape rather than by a single, easily accessible minimum of the free energy (a problem we will encounter again in the next Chapter in the context of protein folding), an alternative approach to determine the secondary structures is desirable.

An alternative to free energy minimization can be based on a dynamic approach, in which the transitions between different minima in the structure landscape is explored directly. This is one step closer to reality, since in the approach we have described before, the real physical formation process of the RNA fold is not at all reflected.

In reality, the folding process occurs via a stochastic opening and closing process of helical segments. Two structures which differ by one helix can be visualized as being separated by a kinetic barrier, shown in a schematic drawing of Figure 3.3. The barrier crossing can be considered as an activated process, similar to chemically activated processes, and hence it is controlled by a rate expression

$$k_\pm = k_0 \cdot \exp -\beta \Delta G_\pm \tag{3.43}$$

where the attempt frequency k_0 for helix nucleation is estimated from experimental data to lie at about $k_0 \approx 10^8 s^{-1}$; it reflects only local stacking processes with a transient nucleation core. The free energy differences $\Delta G_\pm$

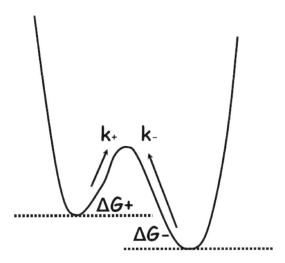

FIGURE II.3.3: Kinetic barrier between two RNA folds, differing by one helix.

in eq.(3.43) correspond to the difference between the transition state (the barrier) and the two local minima.

H. ISAMBERT and E. D. SIGGIA, 2000, have proposed a two-scale model for the RNA free energy to compute these rates. Their model considers the molecule as being built from individual helices, characterized by the corresponding free energies, and an additional conformational entropy of the entire structure, in which pseudoknots are explicitly permitted. Helices are modelled as rods, while the unpaired chains are considered as Gaussian chains with a *Kuhn length* of 1.5 nm (2.5 bases of 6 Å size). The calculation of conformational entropy is based on substructures (termed *nets* by the authors) for which the entropy is calculated exactly, and combining them in a coarse-grained description. Within the helices, the model also takes into account the base stacking energies, and hence allows to treat sequence-dependent effects in an explicit manner.

A program with which the kinetic folding behaviour of RNA can be computed with this appraoch is *Kinefold* (A. XAYAPHOUMMINE et al., 2005.) Some illustrative results of this approach are shown in Color Figure 1 for the sequence of a mini-RNA of the organism *Xenopus tropicalis* (DQ066652); the sequence contains 79 bases. The two configurations with lowest free energy are shown; the difference between the two is 1.7 kcal/mol. Note that both configurations are structurally very different: the lowest free energy configuration has a rather linear structure, while the other configuration contains a pseudoknot. Recently, results from such kinetic simulations have been combined with single molecule experiments on RNA folding kinetics (S. HARLEPP et al., 2003). In these experiments, *E. coli* 16S rRNA was used, which is 1540 bp long, and shown to have a rather well-structured unfolding pathway under mechanical stretching. Experiment and simulation have provided nice direct evidence for the existence of the rugged landscape of RNA folds.

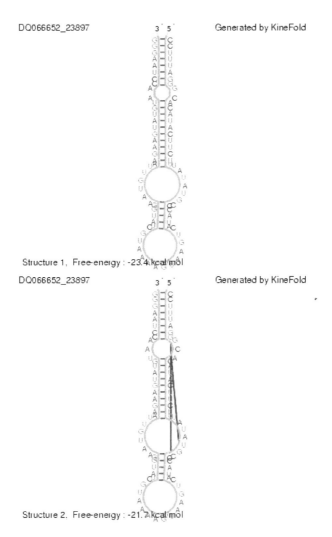

FIGURE 1: RNA structure predicted by the program Kinefold for a 79 bp mini-RNA of Xenopus tropicalis. Direct output of the program, see calculated free energies at the bottom.

FIGURE 2: Two-state folder CI2, the chymotrypsin inhibitor 2. Structural elements are an α-helix and beta sheets (flat arrows). C. MERLO et al.; Copyright (2005) National Academy of Sciences U.S.A.

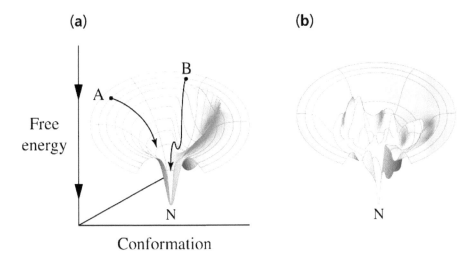

(a)

(b)

Free energy

A

B

N

Conformation

N

FIGURE 3: The folding funnel; a) folding paths from two different initial conditions: A follows a direct path whereas B is trapped in a metastable intermediate; b) rugged energy landscape with multiple minima at intermediate energies. Note that the 'true' equilibrium has a much lower free energy. (Reprinted with permission from Macmillan Publishers Ltd., K. A. DILL and H. S. CHAN, 1997.)

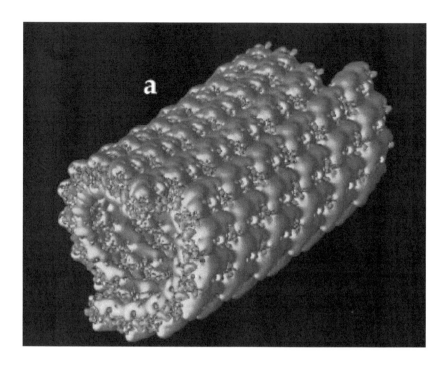

FIGURE 4: The local electrostatic potential of a microtubule. Red and blue values indicate values of the potential of opposite sign. N. A. BAKER et al.; Copyright (2001) National Academy of Sciences U.S.A.

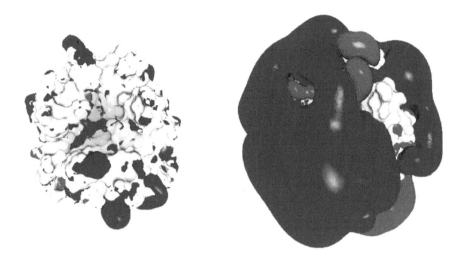

FIGURE 5: Comparison of local (left) and nonlocal (right) electrostatic potential of the enzyme trypsin. The color code indicates postive and negative values of the potential for a selected threshold. Reprinted with permission from A. HILDEBRANDT, Copyright (2005) Rhombos-Verlag.

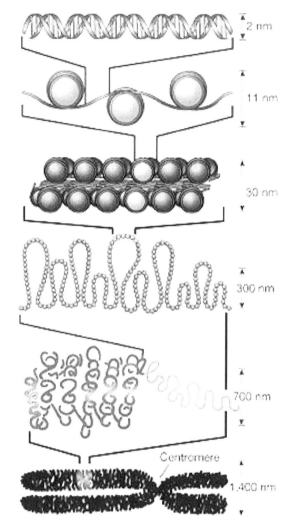

FIGURE 6: Chromatin structure. The base level of compaction is the 11 nm fiber, in which the nucleosomes are positioned as 'beads on a string'; nucleosomal arrays condense to form the 30 nm fiber of yet unknown structure; higher order compactions lead to the familiar chromosome shape of condensed chromatin. (Reprinted with permission from Macmillan Publishers Ltd., G. FELSENFELD and M. GROUDINE, 2003.)

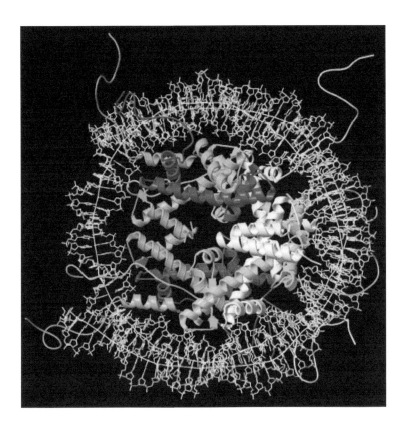

FIGURE 7: Nucleosomal core particle in a ribbon presentation of the core histones (H2A: green; H2B: light green; H3: cyan; H4: red) bound to DNA (yellow). The axis of the wrapped DNA is represented by a white line following the centerline between the base pairs. Notice the dangling histone tails; the first 20 residues of the H3 tails are omitted for clarity. (Reprinted with permission from Macmillan Publishers Ltd., G. FELSENFELD and M. GROUDINE, 2003.)

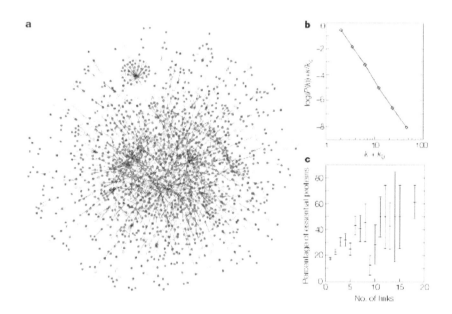

FIGURE 8: a) Protein network graph. Is such a network scale-free? See the range of observed scaling behaviour under b). (Reprinted with permission from Macmillan Publishers Ltd., H. JEONG et al., 2001.)

Additional Notes

In these notes, we remark on three different lines of approaches for the RNA structure prediction problem, which we classify as follows:

Improved folding algorithms. An alternative algorithm for pseudoknot prediction have been proposed by E. RIVAS and S. R. EDDY, 2000. More recent work is by R. M. DIRKS and N. A. PIERCE, 2003, and J. RUAN et al., 2004. The latter work is of particular interest, as it includes *comparative information*, i.e., it uses information from RNA molecules with known secondary structure for the prediction of unknown structures. This - if one likes, 'evolutionary' - approach has been very successful in the original predictions for the structure of ribosomal RNA (G. E. FOX and C. R. WOESE, 1975), and is becoming of renewed recent interest. The inclusion of sequence information can be performed, e.g., based on sequence alignment (D. SANKOFF, 1985). Further recent examples are the works by B. KNUDSEN and J. HEIN, 1999, I. L. HOFACKER et al., 2004, and O. PERRIQUET et al., 2003, H. TOUZET and O. PERRIQUET, 2004.

Finally, we mention a novel approach by R. GIEGERICH et al., 2004, who define an abstract *RNA shape* (as opposed to structure) for a given structural class. The RNA defines the free energy minimum within this class. This construction reduces significantly the set of suboptimal structures that otherwise need to be considered.

More realism. Folding algorithms are continuously improved in order to become more realistic for applications. E.g., ions strongly affect tertiary structure (D. E. DRAPER, 2004). The thermodynamic data underlying the quantitative predictions of RNA structures have been continuously refined and chemical modifications have been included as constraints into the free energy minimization (D. H. MATHEWS et al., 2004). Chemical modifications are a frequently used experimental tool to determine the structure of hypothesized RNA molecules.

Mini-RNA's. The advent of miRNA's has come as a new challenge for the detection of putative RNA's in non-coding regions. I. L. HOFACKER et al., 2004, have developed a technique to perform genome-wide searches for structural elements in time $\mathcal{O}(N \times L^2)$, where N is the genome length, and L the sequence length.

Finally, a standard reference on RNA is the book *The RNA World*, edited by R. F. GASTELAND, T. R. CECH and J. R. ATKINS, 1999.

References

R. Bundschuh and T. Hwa, *Statistical mechanics of secondary structures formed by random RNA sequences*, Phys. Rev. E **65**, 031903 (2002)

R. Bundschuh and T. Hwa, *Phases of the secondary structures of RNA sequences*, Europhys. Lett. **59**, 903-909 (2002)

R. M. Dirks and N. A. Pierce, *A Partition Function Algorithm for Nucleic Acid Secondary Structure Including Pseudoknots*, J. Comp. Chem. **24**, 1664-1677 (2003)

D. E. Draper, *A guide to ions and RNA structure*, RNA **10**, 335-343 (2004)

S. R. Eddy, *How do RNA folding algorithms work?*, Nat. Biotech. **22**, 1457-1458 (2004)

G. E. Fox and C. R. Woese, *5S RNA secondary structure*, Nature **256**, 505-507 (1975)

R. F. Gasteland, T. R. Cech and J. F. Atkins, *The RNA World*, 2nd ed., Cold Spring Harbour Laboratory Press (1999)

R. Giegerich, B. Voß and M. Rehmsmeier, *Abstract shapes of RNA*, Nucl. Acids. Res. **32**, 4843-4851 (2004)

S. Harlepp, T. Marchal, J. Robert, J.-F. Léger, A. Xayaphoummine, H. Isambert and D. Chatenay, *Probing complex RNA structures by mechanical force*, Eur. Phys. J. E. **12**, 605-615 (2003)

I. L. Hofacker, B. Priwitzer and P. F. Stadler, *Prediction of locally stable RNA secondary structures for genome-wide surveys*, Bioinformatics **20**, 186-190 (2003)

H. Isambert and E. D. Siggia, *Modeling RNA folding paths with pseudoknots: Application to hepatitis virus ribozyme*, Proc. Natl. Acad. Sci. USA **97**, 6515-6520 (2000)

B. Knudsen and J. Hein, *RNA secondary structure prediction using stochastic context-free grammars and evolutionary history*, Bioinformatics **15**, 446-454 (1999)

D. H. Mathews, M. D. Disney, J. L. Childs, S. J. Schroeder, M. Zuker and D. H. Turner, *Incorporating chemical modification constraints into a dynamic programming algorithm for prediction of RNA secondary structure*, Proc.

Natl. Acad. Sci. USA **101**, 7287-7292 (2004)

M. L. Mehta, *Random Matrices*, 2nd ed., Academic Press (1991)

R. Nussinov and A. B. Jacobson, *Fast algorithm for predicting the secondary structure of single-stranded RNA*, Proc. Natl. Acad. Sci. USA **77**, 6309-6313 (1980)

H. Orland and A. Zee, *RNA folding and large N Matrix theory*, Nucl. Phys. B **620**, 456-476 (2002)

O. Perriquet, H. Touzet and M. Dauchet, *Finding the common structure shared by two homologous RNAs*, Bioinformatics **19**, 108-116 (2003)

M. Pillsbury, J. A. Taylor, H. Orland and A. Zee, *An Algorithm for RNA pseudoknots*, q-bio/0405014 (2003)

M. Pillsbury, H. Orland and A. Zee, *A steepest descent calculation of RNA pseudoknots*, Phys. Rev. E **72**, 011911 (2005)

E. Rivas and S. R. Eddy, *A dynamic programming algorithm for RNA structure prediction including pseudoknots*, J. Mol. Biol. **285**, 2053-2068 (1999)

J. Ruan, G. D. Stormo and W. Zhang, *An Iterated loop matching approach to the prediction of RNA secondary structures with pseudoknots*, Bioinformatics **20**, 58-66 (2004)

D. Sankoff, *Simultaneous solution of the RNA folding, alignment and protosequence problems*, SIAM J. Appl. Math. **45**, 810-825 (1985)

H. Touzet and O. Perriquet, *CARNAC: folding families of related RNAs*, Nucl. Acids Res. **32**, W142-W145 (2004)

G. Vernizzi, H. Orland and A. Zee, *Enumeration of RNA structures by Matrix Models*, Phys. Rev. Lett. **94**, 168103 (2005)

G. Vernizzi, H. Orland and A. Zee, *Prediction of RNA pseudoknots by Monte Carlo simulations*, condmat/0310505 (2003)

H. Xayaphoummine, T. Bucher and H. Isambert, *Kinefold web server for RNA/DNA folding path and structure prediction including pseudoknots and knots*, Nucl. Acids. Res. **33**, 605-610 (2005)

M. Zuker, *On Finding All Suboptimal Foldings of an RNA Molecule*, Science **244**, 48-52 (1989)

Chapter 4

Proteins

4.1 Proteins: folding

The fact that there are twenty building blocks for a protein - the amino acids - makes the problem of relating sequence to structure even more difficult to solve than for RNA with its four base units, the nucleotides.

The puzzling feature of the protein folding problem is best illustrated by invoking a famous paradox, first stated by Cyrus Levinthal (C. LEVINTHAL, 1969). If one starts out with the protein as a linear polymer and wants to fold it into its 3-D spatial structure which one assumes known from X-ray crystallography, each residue in the chain has about ten times more conformational positions available as in its native state. Thus the total number of conformational shapes for a 100-residue protein is 10^{100}. Supposing a conformational change at the single residue level happens at a picosecond rate, a fully random exploration of the landscape would take 10^{81} years to find the 'right' structure.

Obviously, this is not what happens. Following a metaphor coined by P. WOLYNES, 2001, the random walk of the protein conformations in Levinthal energy landscape resembles that of a golf course:[1] all energy levels are equal, except for the right one.

The consensus on protein folding that has emerged up to now can be summarized as follows:

- 'Small' proteins are two-state folders. This reminds us, for good reason, of the denaturation/hybridization problem of DNA;

- 'Medium' size proteins are 'complex' because *many* details of the composition matter for the final folded state;

- 'Large' proteins do not have the problem we discuss: they simply do not fold spontaneously. They are folded with the help of a specialized

---

[1]This should hold at least for the putting green.

machinery, the chaperones. We will not address this topic here any further.

We begin with the small proteins.

Small proteins. We'll start with the *two-state folders*. Ideally, a two-state folder should have exactly two states: a denatured (coil) state, and the native folded (helix) state. If this is the case, the discussion of Chapter 2 on the denaturation of short chains applies: native and denatured state are describable as a chemical equilibrium; the only difference being the calculation of the free energy. So we are done.

If true, this would imply that the folding of small proteins is about as featureless as the denaturation/hybridization of a short DNA fragment, and only the melting temperature (here: the folding temperature) would depend somewhat on amino acid composition. We would then expect the following under *mutations* of the amino acid composition of the chain: suppose we replace amino acids and look at the folding profile in the same way as we did for the melting profile of DNA. If the amino acids have only different binding strengths, the resulting curves should be shifted in temperature but otherwise not much affected.

But that is not what is observed. Studies of protein unfolding kinetics have uncovered quite a different picture. The effect of mutations on folding can be quantified by the so-called Φ-*value*

$$\Phi \equiv \frac{RT \ln(k_{wt}/k_m)}{\Delta G_M} \, , \tag{4.1}$$

where k_{wt} and k_m are the folding rates of the wild-type and mutated proteins, and ΔG_M is the free energy associated with the change in thermal stability due to the mutation. Φ is also related to the free energy difference between the *transition state* and the native state in the non-mutated protein, ΔG_T. One writes $\Phi = \Delta G_T/\Delta G_M$, neglecting prefactor differences.

Φ is found to have values between 0 and 1, indicating that the residues are in the ensemble of transition states. These are those pseudo-equilibrium structures between the folded or unfolded equilibrium structures which have at least a partial native-like structure. Sometimes values less than 0 or larger than 1 appear which are not so easy to interpret. In fact, the most important observed feature of Φ is that changes in neighbouring bases can have very different effects on the Φ-value. Φ thus does not vary in a continuous manner upon mutations along the chain.

In hindsight this is not surprising, given that the various amino acids have a rather variable chemical nature, in contrast to the nucleic acids.

Let's discuss this issue in more detail, for one particular example, the chymotryspin inhibitor CI2. Its structure is shown in Color Figure 2. We base the discussion on a simple model specifically tuned to this case. The molecule CI2 has one α-helix and a four-stranded β-sheet. In the α-helix, 20 single residue mutations have been studied experimentally with Φ-values in the range between $-0.35 \leq \Phi \leq 1.25$. What does a simple model look like to reproduce these data, at least the underlying trend?

In a very simple ansatz one can attempt to describe the folding kinetics by a master equation (see Part I, Chapter 2)

$$\dot{\mathbf{p}}(t) = \mathbf{W}\mathbf{p}(t), \tag{4.2}$$

where $\mathbf{p}$ is the vector of the state probabilities of the protein. We assume the transition state from a state m to a state n to be given by

$$w_{nm} = \frac{1}{t_0}\left[\frac{1}{1 + exp[(G_n - G_m)/RT]}\right], \tag{4.3}$$

where G_i is the free energy of each partially folded state, and t_0 is a characteristic time.

The solution of eq.(4.2) is given by

$$\mathbf{p}(t) = \sum_\lambda \mathbf{Y}_\lambda \exp[-\lambda(t/t_0)] \tag{4.4}$$

with eigenvectors Y_λ and eigenvalues λ. C. MERLO et al., 2005, assumed the four folding states:

- the denatured state (D);

- a partially folded state (α) with a α-helix;

- a partially folded state (β) with a β-sheet;

- the native state (N).

Hence in this case, the matrix $\mathbf{W}$ is a (4×4)-matrix.

Due to its simplicity, the model can be solved exactly, and gives the eigenvalues

$$\lambda_{0,...,3} = (0, \ 1 - q, \ 1 + q, \ 2) \tag{4.5}$$

with

$$q = \frac{1 - \alpha\beta/N}{\sqrt{(1 + \alpha)(1 + \beta)(1 + \alpha/N)(1 + \beta/N)}} \tag{4.6}$$

with $-1 < q < 1$, and where $\alpha \equiv e^{-G_\alpha/RT}$; analogously for D, N and β.

Despite its four states, the model is found to display two-state kinetics provided the free energy of the native state is much smaller than the free energy of the other states and, further, if the α- and β-intermediaries have free energies larger than D. In this case a kinetic barrier between D and N exists, and the transition is governed by a single-exponential dynamics.

Given these conditions, q in eq.(4.6) can be simplified to (*Exercise*)

$$q \approx ((1 + \alpha)(1 + \beta))^{-1/2} , \tag{4.7}$$

and the folding rate k - i.e., the smallest relaxation rate λ_0 - simplifies upon expansion of the square-root for $\alpha, \beta \ll 1$ to

$$k \equiv 1 - q \approx \frac{\alpha + \beta}{2} . \tag{4.8}$$

From this result one infers that the folding rate is simply the sum of the rates of folding into either the α- and or β-substructures with equal probability.

The model can now be used to study the effect of mutations. Treating the effect of the mutations on the free energies as small perturbations ΔG one can write

$$\ln\left(\frac{k_{wt}}{k_m}\right) \approx \frac{\partial \ln k}{\partial \ln G_\alpha} \Delta G_\alpha \tag{4.9}$$

with k taken from eq.(4.8). Consequently, with $i = \alpha, \beta$,

$$\Phi = \chi_i \frac{\Delta G_i}{\Delta G_N} \tag{4.10}$$

with, e.g., for $i = \alpha$

$$\chi_\alpha = -RT \frac{\partial \ln k}{\partial G_\alpha} = \frac{\alpha}{\alpha + \beta} . \tag{4.11}$$

Within this simple model one finds that Φ is the product of a *structural factor* χ and an *energetic factor* $\Delta G_i/\Delta G_N$. Although χ obviously involves energies, it can nevertheless be considered structural since it explicitly depends on the possible intermediaries, be they the formation of an α-helix or a β-sheet.

Coming back to the protein CI2, the experimental Φ-values of the twenty mutations on the α-helix could indeed be reproduced with this model with a satisfactory correlation coefficient of 0.85 (C. MERLO et al., 2005). It should be kept in mind, however, that the calculation is indeed a simplified one: in reality there may well be interactions (so called *tertiary contacts*) between the α-helix and the β-sheet; the model assumes that both fold independently from each other.

Medium size proteins. Already from the foregoing discussion of the small proteins we have learnt that the folding process follows not only one pathway, but has several options. In fact, it is more reasonable to think of a characterisation of the folding process as an ensemble of possible pathways which can be visualized as trajectories in a complex energy landscape. The energy space of protein conformations is what is sometimes called a 'rugged' landscape, with many minima of near degeneracy - quite similar to what we have encountered already in the case of RNA.

An elegant concept to characterize this landscape is the *funnel*, shown in Color Figure 3 (K. A. DILL and H. S. CHAN, 1997). In this plot, free energy is plotted vs. conformation. The folding process of a protein corresponds to a descent in the funnel. In the beginning, near the top of the funnel, a large number of possible pathways exist for the protein. The more it descends in the funnel, the more limited becomes the number of available pathways: the funnel narrows down towards the conformation with the minimal free energy.

Let's try to quantify this idea a bit. For this we turn to a basic model from statistical physics that rationalizes the idea of a random energy landscape. The *Random Energy Model* (REM) was invented by B. DERRIDA, 1980, with the idea to represent a class of models of disordered systems with many nearly degenerate minima.[2] For its application to protein folding, we follow J. N. ONUCHIC et al., 1997, and consider a simple model with two basic variables, the energy[3] E, and one additional quantity, which we take as the fraction of *native-like contacts*, Q, hence a measure for the structural similarity between a given protein conformation and the native one. For the native structure, we take $Q = 1$. Q serves as our conformational coordinate.

For this model can define a number of useful physical quantities like the thermal average of the energy, $\overline{E}(Q)$, the roughness of the energy landscape (i.e., the fluctuations in energy), $\sqrt{\Delta E^2(Q)}$, and the density of states, $\Omega(E, Q)$. The entropy of a configuration, e.g., is then given by $S(E, Q) = k_B \ln \Omega$.

---

[2]The most prominent example of such systems is the 'spin glass', which is a disordered ferromagnetic system in which a large number of ferromagnetic couplings have been replaced by antiferromagnetic couplings. Since the latter favor the orientation of the neighbouring magnetic moments in antiparallel direction, the system can become 'frustrated' since the spins may underlie conflicting conditions for the orientation of their magnetic moments. This gives rise to a large degeneracy of the microstates of the system.

[3]Which we assume as averaged over the solvent.

One key feature of the REM-model is the assumption of a Gaussian distribution of energy states,

$$P(Q, E) = \frac{1}{\sqrt{2\pi\Delta E^2(Q)}} \exp\left(-\frac{(E - \overline{E}(Q))^2}{2\Delta E^2(Q)}\right). \tag{4.12}$$

If a protein has N residues, the total number of its conformations is

$$\Omega_0 = \gamma^N \tag{4.13}$$

where γ is the number of configurations per residue. This number is amenable to simplifications: e.g., when only the backbone coordinates are taken, $\gamma \approx 5$, while the inclusion of excluded-volume effects allows $\gamma \approx 1.5$.

Since by going down the energy funnel the folded structures become more and more to resemble the native state, the total number of configurations decreases: the native structure has a unique backbone conformation. If one calls $\Omega_0(Q)$ the density of conformational states of measure Q and $S_0(Q)$ the corresponding entropy, then the density of conformations with associated energy E is given by

$$\Omega(Q, E) = \Omega_0 P(Q, E) \tag{4.14}$$

and the total entropy is

$$S(Q, E) = S_0(Q) - k_B \frac{(E - \overline{E}(Q))^2}{2\Delta E^2(Q)}. \tag{4.15}$$

At thermal equilibrium, the most probable energy is given by the maximum of the distribution

$$E_{mp} = \overline{E}(Q) - \frac{\Delta E^2(Q)}{k_B T}, \tag{4.16}$$

and the corresponding values of Ω and S at this maximum are easy to compute using eqs.(4.14), (4.15) (*Exercise*).

Further, we can compute the free energy of a misfolded structure of a given structural similarity Q, and at a given temperature T

$$F(Q, T) = E_{mp} - TS(E_{mp}, Q) = \overline{E}(Q) - \frac{\Delta E^2(Q)}{2k_B T} - TS_0(Q). \tag{4.17}$$

Without precise knowledge about the dependence of this expression on Q, further analysis is not possible. To proceed we consider the simplest case, namely that the free energy has two minima, one at $Q \approx 0$, corresponding to an ensemble of collapsed misfolded states with a varying degree of ordering, and another at the folded state[4], $Q = 1$. As before in the discussion of the small proteins these states will be separated by a kinetic barrier, hence by energetic and entropic contributions. If we neglect the entropy of the folded state to a first approximation, we have

$$F_{native} = E_N .\tag{4.18}$$

At the folding temperature T_F, the free energies of the folded and unfolded state coincide

$$F_{native}(Q = 1, T_F) = F(Q_{min}, T_F)\tag{4.19}$$

where the value of Q_{min} is close to that of the unfolded state. From this we can compute the *stability gap*

$$\delta E_s \equiv \overline{E}(Q_{min}) - E_N = S_0 T_F + \frac{\Delta E^2(Q_{min})}{2k_B T_F} .\tag{4.20}$$

The folding temperature T_F can be related to the *glass transition temperature*, at which the entropy of the system vanishes, $S(E_0, Q) = 0$. It is given by

$$T_G(Q) = \sqrt{\frac{\Delta E^2(Q)}{2k_B S_0(Q)}} ,\tag{4.21}$$

with $S_0 = k_B \ln \Omega_0$ with Ω_0 given before.

The transition into the glassy state occurs precisely when there are too few states available, and the system remains frozen in one of those states. The ratio of the folding to the glass temperature is approximately given by

$$\frac{T_F}{T_G} \approx \frac{\delta E_s}{\sqrt{\Delta E^2}} \cdot \sqrt{\frac{2k_B}{S_0}} .\tag{4.22}$$

In order for a protein to fold correctly and not to be caught frozen in a wrong minimum, one needs to fulfill the condition

$$T_F > T_G .\tag{4.23}$$

Since S_0, ΔE^2 and E_N all depend linearly on N, the ratio eq.(4.22) is not dependent on protein length, but very sensitive to interaction energies. This explains why the folding process remains so difficult to capture quantitatively.

Beyond this basic picture given here, there is still no general consensus on how to quantitatively model and predict protein folding; maybe there is

---

[4]The discussion of the small proteins indicates how to extend this idea to more available states.

no such theory ever to have, since such a great number of details intervene: proteins are individuals, designed for very specific purposes, and this individuality may also be reflected in the way they fold. In the Additional Notes some interesting novel conceptual avenues to attack the folding problem are listed. The problem is likely to stay with us for some more time. Here, we now move on to the problem of protein docking.

4.2   Proteins: docking

The protein docking problem as a specific bioinformatics problem dates back to the late seventies (J. JANIN and S. J. WODAK, 1978) and has since seen a continuous development towards more efficient algorithms. The term protein docking encompasses two closely related situations: *protein-protein docking*, as it occurs during the formation of protein complexes, and *protein-ligand docking*, which relates to the drug design problem mainly driving the field. Here, protein docking is an essential tool to help design specific and efficient small drug molecules.

In this section, we discuss the two main statistical physics aspects of the problem. In order to fit two molecules snugly together, they must certainly fit geometrically; but also their interactions must be compatible. While in fact both geometry and interaction are fundamentally coupled, the approximate approaches that exist to the docking problem today usually separate these questions in two independent procedures. *Geometric fit* is used as a first level of screening for docking candidates, and only then is the quality of the docking evaluated ('scored') by a quantitative measure of the physical interaction of the docking partners.

Geometric fit. In order to determine the geometrical fit of two molecules, a quantitative method to determine surface complementarity is needed. We explain the algorithm introduced in (E. KATCHALSKI-KATZIR et al., 1992). It is based on the projection of the molecules in their atomic coordinates on a 3-dimensional grid built from N^3 grid points. The molecules are represented by discrete coordinates

$$a_{\mathbf{k}}^i = \begin{cases} 1 & \epsilon\ \Gamma \\ \varrho^i & \epsilon\ \Omega \\ 0 & \epsilon\ \Sigma \end{cases} \tag{4.24}$$

where $i = 1, 2$ represents the molecule of volume Ω and surface Γ, $\mathbf{k} = (l, m, n)$ are the lattice indices. A lattice point is considered inside the molecular volume Ω if there is at least one atom within a distance r from it, where r is a parameter on the order of one *van der Waals atomic radius* - a measure for the short-range repulsion of the electron shells. The surface layer Γ then is a finite layer between the inside of the molecule and its exterior, Σ.

Matching of the surfaces is achieved by calculating correlation functions. The correlation between the a^i is given by

$$c_\mathbf{q} = \sum_{k=1}^{N} a_\mathbf{k}^1 \cdot a_\mathbf{k+q}^2 \tag{4.25}$$

where $\mathbf{q}$ denotes the number of grid steps by which molecule a^2 is shifted with respect to molecule a^1. If the molecules have no contact, $c_\mathbf{q} = 0$. In order to penalize interpenetration of the molecules, ϱ^i will be assigned a large negative value for $i = 1$, and a small positive value for $i = 2$, leading to an overall negative contribution in $c_\mathbf{q}$. In this way, positive and negative correlations can clearly be related to the geometric positioning of the molecules.

Computationally, the determination of $c_\mathbf{q}$ would require the calculation of N^3 multiplications and additions for each of the N^3 relative shifts $\mathbf{q}$, i.e. N^6 steps. This can be reduced by taking advantage of the lattice representation. Calculating first the discrete Fourier transform of the $a_\mathbf{q}^i$, which can be done with the Fast Fourier Transform (FFT), and then going back by an inverse Fourier transform, the number of computational steps can be reduced to at most $\mathcal{O}(N^3 \ln N)$.

Docking energetics. The molecules typically have to find their partners in solution. As a consequence, the first quantity to determine is the *free energy of solvation* for each individual molecule. The Gibbs free energy for this process can be written as

$$\Delta G \equiv G^{solvent} - G^{vacuum} = \Delta G^{polar} + \Delta G^{nonpolar} . \tag{4.26}$$

The nonpolar contribution consists of various terms, which are usually ascribed to cavity formation (i.e., to account for the cost of digging a hole into water to fit in the protein), conformational contributions, and van der Waals interactions. It is very difficult and debated how these quantities can be modelled; in any case, it is an approximate procedure. Here, we will be interested in electrostatic interactions only, which dominates the polar part in eq.(4.26). For the solvation process shown in Figure 4.1, the relevant quantity is the *free energy of binding*. It is given by (R. M. JACKSON and M. J. E. STERNBERG, 1995)

$$\Delta G^{bind} = \Delta\Delta G^{solv,A} + \Delta\Delta G^{solv,B} + \Delta\Delta G^{solv,A-B} \tag{4.27}$$

The three contributions arise from: i) the change in solvation energy of molecule A upon binding; ii) the change in solvation energy of molecule B

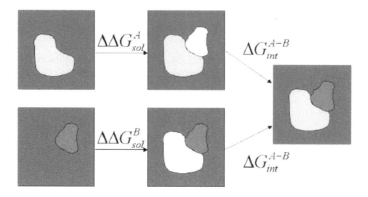

FIGURE II.4.1:   Solvation of a protein complex in water; for explanation, see text. Reprinted with permission from A. HILDEBRANDT, 2005; Copyright Rhombos-Verlag.

upon binding; iii) the interaction energy of A and B in the presence of solvent. The last contribution contains the energy stored in the bonds between both proteins, and in principle both polar and non-polar contributions.

4.3   Electrostatics

The Poisson-Boltzmann equation. In this section we will see how to describe charge effects in biomolecules, provided some simplifying assumptions can be made. Electrostatic effects are very important in biomolecules, since most of them are highly charged, like DNA and proteins. Charge effects are indeed so important that they even play an important role in mechanisms of gene regulation in eukaryotes, since they are employed to condense the DNA in the cellular nucleus. We will discuss this issue in some detail at the end of this chapter, but first we have to set up the basics.

The assumptions on which we will base the discussion of electrostatic effects of biomolecules are, for a start:

- We consider only direct charge interactions and ignore permanent and induced dipolar interactions or other;

- All charges are considered as point charges, there are no short-range force effects;

- Water is assumed as a continuous medium with a dielectric constant of $\epsilon \approx 80$.

All of these assumptions are made, in one way or another, in an attempt to separate length scales. We will now give a short-list of the length-scales that are the most common in biomolecular electrostatics.

The Bjerrum length. Within our simplified approach only two energies appear, the Coulomb energy between the charges, and the thermal energy. Their balance gives rise to the definition of the *Bjerrum length*[5]

$$\ell_B \equiv \frac{e^2}{\epsilon k_B T} \tag{4.29}$$

which for $\epsilon = 80$ yields a quantitative value of $\ell_B = 7$ Å. The physical interpretation of this length scale is simple: for two oppositely charged particles at a distance $r < \ell_B$, electrostatics wins, and the particles are bound, while for $r > \ell_B$ thermal fluctuations make the particles unbind.

We can put this simple picture on a more solid footing. The basic equation of classical electrostatics is the *Poisson equation* for the electrostatic potential,

$$\epsilon \nabla^2 \phi(\mathbf{x}) = -4\pi \varrho(\mathbf{x}) \tag{4.30}$$

where the charge density $\varrho(\mathbf{x})$ is given by

$$\varrho(\mathbf{x}) = \sum_i Z_i e c_i(\mathbf{x}) + \varrho_{ext}(\mathbf{x}) \,. \tag{4.31}$$

The first term describes mobile ion charges of type i and charge Z_i in the solvent, while the second term is given by a fixed charge distribution. The mobile charges are distributed according to a Boltzmann weight

$$c_i(\mathbf{x}) = c_{0,i} e^{-Z_i e \phi(\mathbf{x})/k_B T} \,. \tag{4.32}$$

---

[5] A note on units. In the discussion of biomolecular electrostatics, we use the cgs-system, and not the SI-system (and override hereby the opinion of a reader of an early draft of this book who claimed that everyone younger than 35 hates cgs units. Sorry, Martin!). In my view cgs-units do not disfigure the mathematical expressions as much as SI units do, but this becomes a real problem only when magnetic properties are addressed. Since in the discussion of electrostatics we only need the Poisson equation and little more, the translation from cgs to SI is easy. If we write Maxwell's equation in vacuum as

$$\nabla \cdot \mathbf{E} = -4\pi k \varrho \tag{4.28}$$

we get the cgs-expression with $k = 1$ and the SI-expression with $k = 1/(4\pi\epsilon_0)$. For the mathematically inclined, the choice of $k = 1/(4\pi)$ removes all units, following HEAVISIDE. Finally, the standard book on classical electrodynamics by J. D. JACKSON contains a detailed translation manual.

For the case of an electrolyte solution of two charge types with $Z = \pm 1$, and without external field, the *Poisson-Boltzmann equation* reads

$$\nabla^2 \varphi(\mathbf{x}) = 8\pi \ell_B Z c_s \sinh(Z\varphi(\mathbf{x})) \tag{4.33}$$

with c_s as the bulk ion density. Note that we have rescaled the electrostatic field by $\varphi = e\phi/k_B T$.

Debye screening length. As we see from eq.(4.33), the Poisson-Boltzmann equation is a nonlinear equation. Under certain conditions (see below) it can be linearized, yielding the *Debye-Hückel equation*

$$\nabla^2 \varphi(\mathbf{x}) = \kappa^2 \varphi(\mathbf{x}) \tag{4.34}$$

with

$$\kappa^{-2} = \frac{\epsilon k_B T}{8\pi e^2 Z^2 c_s} \equiv \ell_D^2 \tag{4.35}$$

where ℓ_D is the *Debye screening length*. It is ion-density dependent, and varies from a value of 3 Å in 1 M NaCl to about 1 micron in pure water.

The validity of the Debye-Hückel approximation can be determined from a comparison of kinetic and interaction energies of the involved particles. If one takes $c_s^{-1/3}$ as a length describing mean particle separation, then the condition

$$\Gamma \equiv \ell_B c_s^{1/3} \ll 1 \tag{4.36}$$

needs to be fulfilled for the DH-approximation to be applicable.

Counterions at a charged planar surface: the Guoy-Chapman length. As an exemplary application of the nonlinear Poisson-Boltzmann equation we consider the case of counterions, assumed as anions, opposite to a planar surface of negative charge. The latter is described by a surface charge density $\sigma < 0$. Since the system is translationally invariant, the Poisson-Boltzmann equation can be considered in one dimension, orthogonal to the wall in direction z, which yields

$$\varphi''(z) = -(4\pi \ell_B c_s) e^{-\varphi(z)}. \tag{4.37}$$

This equation is complemented by the boundary condition

$$\left. \frac{d\varphi}{dz} \right|_{z=0} = -\frac{4\pi}{e}\sigma > 0. \tag{4.38}$$

Upon integration one finds

$$\varphi(z) = 2\ln(z + \ell_{CG}) + \varphi_0 \tag{4.39}$$

where the constant potential φ_0 is left unspecified for now. The length ℓ_{CG} introduced here is the *Guoy-Chapman length*

$$\ell_{CG} \equiv \frac{\epsilon k_B T}{2\pi e |\sigma|} = \frac{e}{2\pi |\sigma| \ell_B} \sim \sigma^{-1} \tag{4.40}$$

which depends on the surface charge density. The density profile of the mobile charges

$$c_s(z) = \frac{1}{2\pi \ell_B} \frac{1}{(z + \ell_{CG})^2} \tag{4.41}$$

is found to decay algebraically for large distances z, while the potential itself has a logarithmic (hence weak) divergence. The Guoy-Chapman length characterizes the charge density layer at small distances, $z \ll \ell_{CG}$.

Manning condensation. So far, our elementary reasonings allowed us to introduce a number of relevant physical length scales to gain an intuitive idea of the importance of electrostatic phenomena in solution. We now turn to a first biologically motivated application of the Poisson-Boltzmann equation. We consider the electrostatic profile of counterions around a straight cylinder of radius R. This setting can be understood as a highly idealized model for a linear, charged biomolecule like DNA or a polypeptide chain.

For this geometry, the Poisson-Boltzmann equation reads in cylindrical coordinates

$$\frac{d^2\varphi}{dr^2} + \frac{1}{r}\frac{d\varphi}{dr} + \kappa^2 e^{-\varphi(r)} = \frac{\ell_B \sigma}{Ze}\delta(r - R). \tag{4.42}$$

This differential equation eq.(4.42) is of *Liouville-type*, and admits an analytical solution. With the change of variable $x = R\ln(r/R)$ for $r > R$ it is transformed into

$$\frac{d^2\varphi}{dx^2} + \kappa^2 e^{-\varphi(x)+2x/R} = 0. \tag{4.43}$$

The shifted potential $\tilde{\varphi} = \varphi(x) - 2(x/R)$ satisfies the planar Poisson-Boltzmann equation, albeit with a different boundary condition

$$\left.\frac{d\tilde{\varphi}(x)}{dx}\right|_{x=0} = \frac{\ell_B \sigma}{Ze} - \frac{2}{R}. \tag{4.44}$$

The solution of the Poisson-Boltzmann equation is, by analogy to the planar case, given by

$$\tilde{\varphi} = 2\ln\left(1 + \frac{\kappa x}{\sqrt{2}}\right), \tag{4.45}$$

where the boundary condition fixes the value of

$$\kappa = \frac{1}{2}\left(\frac{\ell_B \sigma}{Ze} - \frac{2}{R}\right). \tag{4.46}$$

This result makes no sense for $\ell_B R\sigma < 2Ze$, i.e. $\kappa < 0$. We then have to take the solution of the Poisson-Boltzmann equation for $\kappa = 0$, and obtain instead a logarithmic potential

$$\varphi(r) = 2\xi_m \ln(r/R) \tag{4.47}$$

where

$$\xi_m \equiv \frac{\ell_B R\sigma}{2Ze} \tag{4.48}$$

is the *Manning parameter*. We thus obtain the full solution as

$$\varphi(r) = \begin{cases} 2\xi_m \ln(r/R), & \xi_m \le 1, \\ 2\ln(r/R) + 2\ln[1 + (\xi_m - 1)\ln(r/R)], & \xi_m > 1. \end{cases} \tag{4.49}$$

For $\xi_m > 1$, the electrostatic potential behaves like $\varphi(r) \sim 2\ln(r/R)$, and is essentially independent of charge density; the counterions are bound to the cylinder. For $\xi_m \le 1$, the number of counterions $c_s(r) \sim exp - \varphi(r)$ contained in a cylindrical shell R_0 around the cylinder behaves as

$$c_s(r) \sim r^{2(1-\xi_m)}\Big|_R^{R_0} \tag{4.50}$$

and clearly diverges with R_0: the counterions escape to infinity. The phenomenon of the counterion confinement to the DNA is called *Manning condensation* (G. S. MANNING, 1969).

Protein electrostatics. We will now go one step further and apply our knowledge to proteins, in particular to the determination of solvation free energies of atoms, ions and biomolecules. In many cases, the dominating contribution is due to electrostatics (B. HONIG and A. NICHOLS, 1995), and the main task then is to compute the electrostatic potential of the proteins and their complexes.

The electrostatics problem associated with proteins can be determined by solving the Poisson or Poisson-Boltzmann equation for the geometry depicted in Figure 4.2. It is assumed that the space Ω is filled with an arrangement of charges (not shown explicitly), while we ignore the distribution of ions that can usually be assumed to surround the protein. We thus 'only' have to solve the Poisson equation in this inhomogeneous space. The inclusion of the mobile charges is, of course, possible.

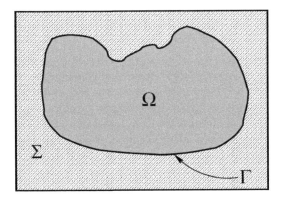

FIGURE II.4.2: Electrostatic problem of a protein in a solvent: a body Ω imbedded in space Σ; Γ defines the boundary. Reprinted with permission from A. HILDEBRANDT et al.; Copyright (2004) American Physical Society.

The mathematical formulation of the problem consequently is:

$$\Delta\phi_\Omega(\mathbf{r}) = -\frac{4\pi}{\epsilon_\Omega}\varrho \tag{4.51}$$

$$\nabla_{\mathbf{r}}[\epsilon_\Sigma(r)\nabla_r\phi_\Sigma(\mathbf{r})] = 0 \tag{4.52}$$

$$\partial_n[\epsilon_\Omega\phi_\Omega - \epsilon_\Sigma\phi_\Sigma]|_\Gamma = 0 \tag{4.53}$$

$$\phi_\Omega|_\Gamma = \phi_\Sigma|_\Gamma \tag{4.54}$$

For proteins, $\epsilon_\Omega \approx 2-5$ is usually assumed, while for the surrounding water, $\epsilon_\Sigma \approx 80$. In the Poisson equation for the solvent we have allowed for a (local) spatial dependence of ϵ_Σ; this is frequently done in a phenomenological way to mimic effects of the modification of the water dielectric properties near the protein. This effect is based on the restricted motion of the water near the protein, bringing about distinct *orientational polarization correlations*. We will see how this effect of water structure can be built into the theory of electrostatics in a more fundamental way in the next subsection.

The solution of the above equations can be obtained only numerically for the complex protein surfaces. Several sophisticated solvers exist for this purpose meanwhile; one of the most advanced examples come from the Holst group which was used to calculate the electrostatic potentials of large biomolecular structures, such as microtubules and ribosomes (N. BAKER et al., 2001), see Color Figure 4.

Nonlocal electrostatics. If we want to describe the electrostatic properties of biomolecules on length scales for which a membrane can be modelled as a fluctuating sheet, a DNA molecule as a cylinder, and a globular protein as a sphere, the level of description of electrostatic interactions is well justified. For the large protein structures we have just seen, which are cylindrical on a mesoscopic scale, but have a lot of atomistic structural detail which may - or will - be of biological relevance, we have been pushing things a bit too far. It seems hard to justify why the water surrounding a protein surface can, on this atomic scale, be considered a dielectric medium of dielectric constant $\epsilon = 80$.

On such length scales, the usual continuum approach ultimately must break down, since water is not featureless on these scales: water has structure. The water molecules respond to the presence of charges, and their network of hydrogen bond has to rearrange, leading to correlations of the water molecule orientations over a characteristic length scale which we denote as λ.

One way to 'repair' the error made in classical electrostatics is to modify the dielectric function. Conventionally this is done by an inclusion of a local spatial dependence $\epsilon(r)$ near the protein surface. This function, which can be parametrized in various ways, is used to effectively reduce the dielectric constant of water to a small value near the protein surface. This approach is, while physically justifiable, a technically uncontrolled procedure.

But there is, at least at present, no real practical alternative to continuum electrostatics as it still is much more computationally efficient than microscopic simulations based on, e.g., molecular dynamics (MD) (T. SIMONSON, 2001). Interest has therefore risen in systematic extensions of the theory of continuum electrostatics that allow to account for spatial variations of the dielectric behaviour of the solvent, in particular near the boundary of a protein.

Within the continuum theory of the electrodynamics of matter, the orientational polarizability of water around a protein is nothing but a spatial dispersion effect. Such effects are known to be tractable within electrodynamic theory, and indeed they have taken into account in an approach called 'nonlocal electrostatics'[6] (A. A. KORNYSHEV et al., 1978; M. A. VOROTYNTSEV, 1978; A. A. KORNYSHEV and M. A. VOROTYNTSEV, 1979) which generalizes

---

[6]The name 'nonlocal electrostatics' is unfortunate since it may create confusion by alluding to strange effects of 'action at a distance' and the like. The nonlocality we talk about is nothing of that sort, it just means that the dielectric response requires to take into account field values not at local points, but over a certain spatial range. All physical fields (potentials, electrostatic fields) remain well-defined local objects.

the commonly used electrodynamics of continuous media to take into account such spatial dispersion effects.

The basis of this theory is the linear relationship between the dielectric displacement field $\mathbf{D}$ and the electric field $\mathbf{E}$ through a permittivity kernel which, in general, depends on two spatial arguments,

$$\mathbf{D}(\mathbf{r}) = \frac{1}{4\pi} \int d\mathbf{r}' \epsilon(\mathbf{r}, \mathbf{r}') \mathbf{E}(\mathbf{r}') . \qquad (4.55)$$

Here, $\epsilon(\mathbf{r}, \mathbf{r}')$ is the dielectric permittivity tensor. It carries the new characteristic length scale, the correlation length λ of the water *orientational polarization correlations* introduced before. This length defines the scale for the deviation of the dielectric properties of a solvent from its average bulk value.

For the protein geometry of Figure 4.3, the generalization of the Poisson equation in the solvent in nonlocal electrostatics reads as

$$\nabla \int_{\Sigma} d\mathbf{r}' \epsilon_{\Sigma}(\mathbf{r}, \mathbf{r}') \nabla' \phi_{\Sigma}(\mathbf{r}') = 0 \qquad (4.56)$$

where the primed symbol ∇' denotes differentiation with respect to $\mathbf{r}'$. The main ingredient of the nonlocal theory is the integral kernel $\epsilon(\mathbf{r}, \mathbf{r}')$ which contains the dependence on the correlation length λ. The mathematical expression for this model depends on the water model one wants to use. A simple, standard example for this quantity is the *Lorentzian model*[7]

$$\epsilon(\mathbf{r} - \mathbf{r}') = \epsilon_{\infty} \delta(\mathbf{r} - \mathbf{r}') + \frac{\epsilon_{\Sigma} - \epsilon_{\infty}}{4\pi\lambda^2} \frac{e^{-\frac{|\mathbf{r} - \mathbf{r}'|}{\lambda}}}{|\mathbf{r} - \mathbf{r}'|} . \qquad (4.57)$$

As can be observed for this example - indeed this turns out to be of rather general nature - one can write eq.(4.57) in a more general form as

$$\epsilon(\mathbf{r} - \mathbf{r}') = \epsilon_{\infty} \delta(\mathbf{r} - \mathbf{r}') + \tilde{\epsilon} \mathcal{G}(\mathbf{r} - \mathbf{r}') \qquad (4.58)$$

where $\tilde{\epsilon} \equiv (\epsilon_{\Sigma} - \epsilon_{\infty})/4\pi\lambda^2$, and $\mathcal{G}$ is a Green function satisfying

$$\mathcal{L}\mathcal{G} = -\delta(\mathbf{r} - \mathbf{r}') \qquad (4.59)$$

with, in the given case, $\mathcal{L} \equiv \Delta - \lambda^{-2}$, and $\mathcal{G}$ as the Green function of the Yukawa potential. Eq.(4.58) contains two terms; the first reduces to the local limit of the dielectric function ϵ_{∞} at small distances $\mathbf{r} \to \mathbf{r}'$ and is cancelled by the term $\propto \epsilon_{\infty}$ in the second term. The remaining contribution is the local

---

[7]The Lorentzian model assumes isotropy of space as an additional simplifying assumption. Although this assumption does not hold strictly, it is an acceptable first approximation.

limit for large distances, ϵ_Σ, the usual macroscopic dielectric constant.

Eq.(4.58) can be taken as the starting point to reformulate the theory of nonlocal electrostatics as a local theory. For this, we in addition represent the dielectric displacement field as a sum of an irrotational part and a solenoidal part (a so-called *Helmholtz decomposition*),

$$\mathbf{D}(\mathbf{r}) = -\nabla\psi(\mathbf{r}) + \nabla \times \xi(\mathbf{r})\,. \tag{4.60}$$

One can show that the solenoidal part, although it explicitly appears in the expression of the dielectric displacement field, does not appear in the equations of the electrostatic potentials (A. HILDEBRANDT, 2005).

We are then finally left with the following system of equations for the electrostatic potentials ϕ and ψ, (A. HILDEBRANDT et al., 2004)

$$\Delta\phi_\Omega = -\frac{4\pi}{\epsilon_\Omega}\varrho \tag{4.61}$$

$$\Delta\psi_\Sigma = 0 \tag{4.62}$$

$$\epsilon_\Omega\,\partial_n\phi_\Omega|_\Gamma = \partial_n\psi_\Sigma|_\Gamma \tag{4.63}$$

$$\phi_\Omega|_\Gamma = \phi_\Sigma|_\Gamma \tag{4.64}$$

$$[\epsilon_\infty\mathcal{L} - \tilde{\epsilon}]\,\phi_\Sigma = 4\pi\mathcal{L}\psi_\Sigma(\mathbf{r})\,. \tag{4.65}$$

Within its reformulation in terms of the local fields ϕ and ψ, the theory of nonlocal electrostatics can now be treated with standard approaches in order to numerically solve them with boundary element methods.

Such a result is shown in Color Figure 5. In this figure, the local and the nonlocal electrostatic potentials of the enzyme trypsin are shown for comparison. It can clearly be seen that the structure of the electrostatic potential obtained in the nonlocal description deviates significantly from the local one which, for the same threshold value of the potential surfaces, hardly reaches beyond the geometric structure of the protein surface.

From this application we can deduce that the electrostatic potential of proteins on Ångstrom scales is markedly influenced by the water properties, and that electrostatic effects are important for the 'visibility' of a protein to its interaction partners in solution.

4.4 Chromatin

Chromatin structure. In this section we break out of the central dogma of molecular biology, DNA → RNA → protein, which has so far been our guiding line. In any organism all these molecules have to act in concert, and the linear sequence should thus be better represented by a diagram with multiple feedback loops.

In order to express genes, they first have to be switched on. This is done by specific proteins, the *transcription factors* in prokaryotes - and rather *transcription factories*, in eukaryotes. The function of these molecules requires recognition capability: proteins can detect their binding sites with high specificity. We will come back to this in Part III of the book when we look at the dynamics of transcription factor binding on DNA.

In eukaryotes an additional problem arises, however. DNA is condensed and packed in the cellular nucleus, making a simple DNA-protein recognition process based on a diffusional search impossible. There must be a way for a protein or enzyme to find out where it has to go in the first place, even when the DNA is still condensed. Or, DNA has to be unpacked only partially, such that the search time for transcription factors becomes reasonably short.

In order to understand the regulation of transcription in eukaryotes, we first have to get an idea of the compact form of DNA in the cell. The condensed form of DNA, in which it is actually a highly organized DNA-protein complex, is called *chromatin*. Chromatin is the structural basis of chromosomes. The different levels of organisation which are at present only partially understood. Our current knowledge is summarized in Color Figure 6 (from G. FELSENFELD and M. GROUDINE, 2003).

The best understood element of chromatin structure is the *nucleosome*. A nucleosome is a DNA-protein complex which consists of an octameric protein complex built out of eight *histone proteins* (of four different kinds), around which DNA is wrapped along 146 bp with 1.75 left-winded turns. A *linker histone* at the exit of the DNA histones completes the structure; the precise positioning of the linker histone is not yet known. The molecular structure of the so-called *nucleosomal core particle* - the DNA wrapped around the histone octamer - has meanwhile been spatially resolved down to 1.9 Å. A ribbon representation of a nucleosomal particle is shown in Color Figure 7 (from G. FELSENFELD and M. GROUDINE, 2003, after K. LUGER at al., 1997).

Unfortunately, the crystal structure of a nucleosome does not tell its full story. The N-terminal tails of the histone proteins are largely unstructured random coils, hence they do not crystallize. These tails are functionally important in two ways:

- The tails are positively charged and can interact with the negatively charged DNA to form the condensed structures of chromatin, leading to both local or global structural changes of chromatin structure;

- The tails are subject to various chemical modifications, brought about by specific enzymes. Some of these modifications can alter the charge of the chain. At the same time, many of the modifications can be read by specific proteins which affect chromatin structure locally. It has been argued that the histone modifications constitute a higher level regulatory code (B. D. STRAHL and C. D. ALLIS, 2000).

These two mechanisms are clearly not fully independent from each other, and are generally referred to as particular examples of *chromatin remodelling*. The notion of remodelling stands for the totality of dynamic structural changes chromatin can undergo and which are relevant for the regulation of transcription.

Histone tail modifications. Enzymatic histone tail modifications are amino acid specific: not only the chemical nature of the amino acid is a determinant, but also its position on the tail. This is indicated in Figure 4.3. Several different types of modifications are presently known. The most important are:

- *Acetylation.* One or several acetyl groups (CH_3CO) are transferred to lysine residues; the addition of one lysine reduces the positive charge of the tail by one. The modification is reversible, i.e., acetylating and deacetylating enzymes have been found.

- *Methylation.* One (or more) methyl groups (CH_3) are added to lysine residues; there is no effect on the charge. The reversibility of this modification is still under scrutiny; the first demethylating enzymes have recently been found, as discussed in A. BANNISTER and T. KOUZARIDES, 2005.

- *Phosphorylation.* This modification can affect several residues: serine, threonine, tyrosine on the one hand, and lysine, histidine, and arginine on the other. The modifications are chemically different for both groups of amino acids.

```
            p     ac
H2A  N - S G R G K Q G G K A R A K A K T R S S R A G L
              5         10      15      20

        rib  ac          ac  ac        ac              p
H2B  N - P E P S K S A P A P K K G S K K A I T K A Q K K D G K K R K R S R K
              5         10      15      20      25      30

              m         m/ac p     ac    ac        ac    m p           m
H3   N - A R T K Q T A R K  S T G G K A P R K Q L A T K A A R K S A P A T G G V K K
                  5         10      15      20      25      30      35

          p     ac  ac        ac        ac        m
H4   N - S G R G K G G K G L G K G G A K R H R K V L R D
              5         10      15      20
```

FIGURE II.4.3: Specific enzymatic modifications on histone tails. (rib = ribosylation, is a rare modification.)

In general, histone tail modifications do differ somewhat between organisms, despite the high evolutionary conservation of the histones, their tails included. While the experimental evidence is growing that the histone tail modifications are read by a specific transcription machinery (T. AGALIOTI, G. CHEN and D. THANOS, 2002), and apparently different modifications do not act independently (W. FISCHLE, Y. WANG and C. D. ALLIS, 2003), it is clear that the underlying 'background' mechanism is based on an electrostatic attraction of the negatively charged DNA and the positively charged histones. It is thus of interest to devise a simplified model view of this compaction mechanism.

Chromatin fiber electrostatics. We here follow H. SCHIESSEL, 2002. He considered the electrostatic interaction between two DNA strands at the entry-exit point of the DNA wrapped around the nucleosome, see Figure 4.4. The angle α between the two DNA strands defines the *entry-exit angle* at the nucleosome. For the geometry of Figure 4.4, it is defined by

$$h'(\infty) \equiv \tan(\alpha/2). \tag{4.66}$$

The relevant question to be answered is that of the dependence of the angle α on salt concentration c_s, or on other charge-affected quantities. Salt contributions are relevant for the screening of electrostatic interactions, since the Debye screening length behaves as

$$\kappa^{-1} = \ell_D = (8\pi c_s \ell_B)^{-1/2} \tag{4.67}$$

where ℓ_B is the Bjerrum length, the measure for the respective importance of

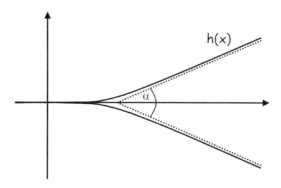

FIGURE II.4.4:   Schematic drawing of the entry-exit region around a nucleosome, after H. SCHIESSEL, 2002. The two bent lines represent the lower and upper DNA strands; the entry-exit angle α is defined via the indicated asymptotics.

electrostatic forces and thermal energy we had introduced before in eq.(4.29).

The DNA strands are considered as two semi-flexible polymers at positions $\pm h(x)$ with persistence length ℓ_P and a *line-charge density* λ. Their free energy is approximately given by

$$F \approx 2k_b T \int_0^\infty dx \left[\frac{\ell_P}{2}(h'')^2 + \ell_B\lambda^2 K_0(2\kappa h(x)) \right] \tag{4.68}$$

where $K_0(x)$ is the *modified Bessel function* of zero order.[8]   The first term in this expression is the bending energy of the strands, while the second describes their electrostatic interaction. The form of this term applies if it can be assumed that the interaction can be considered as that of a straight chain (hence the function K_0) and a single charge at distance $2h$. This approximation works as long as α is not too large; it tends to underestimate the true value.

---

[8] K_α is defined by

$$K_\alpha(x) = \frac{\pi}{2} i^{\alpha+1} H_\alpha^{(1)}(ix)$$

where

$$H_\alpha^{(1)}(x) = J_\alpha(x) + iY_\alpha(x).$$

J_α is the *Bessel function* we have introduced before; the *Neumann function* $Y_\alpha(x)$ is defined by

$$Y_\alpha(x) = \frac{J_\alpha(x)\cos\alpha\pi - J_{-\alpha}(x)}{\sin\alpha x}.$$

The Euler-Lagrange equation for $h(x)$ reads, using the property $K_0'(x) = -K_1(x)$ of the Bessel function,

$$\ell_P h'''' = 2\ell_B \lambda^2 \kappa K_1(2\kappa h) \tag{4.69}$$

together with the boundary conditions

$$h(0) = h'(0) = 0 = h''(\infty) = h'''(\infty). \tag{4.70}$$

A variable change $\tilde{h} = 2\kappa h$, $\tilde{x} = ((4\ell_B \lambda^2 \kappa^2)/l_P)^{1/4} x$ allows to express eq. (4.69) as $h'''' = K_1(h)$ (dropping tildes). With the relation $s = h'(x)|_{x=\infty}$ one immediately obtains the dependence of the opening angle α on the physical parameters, in particular the length scales we defined in Section 4.3,

$$\tan(\alpha/2) = \frac{s}{\sqrt{2}} \left(\frac{\ell_B}{\ell_P}\right)^{1/4} \left(\frac{\lambda}{\kappa}\right)^{1/2}. \tag{4.71}$$

The value of the constant s in this expression is of $O(1)$ (H. SCHIESSEL, 2002).

Solving for α, one finds the dependence

$$\alpha = \arctan(C\, c_s^{-1/4}). \tag{4.72}$$

Intramolecular electrostatic effects can also be included in the calculation. In this case one has to consider the modification of the persistence length ℓ_P by the charges along the chain: due to their mutual repulsion the charges along the molecule increase the rigidity of the chain. This additional repulsion can be calculated and leads to a modification of the persistence length ℓ_P, which has to be replaced by the *Odijk-Skolnick-Fixman persistence length*,

$$\ell = \ell_P + \ell_{OSF} = \ell_P + \ell_B \lambda^2/(4\kappa^2). \tag{4.73}$$

Plugging in numbers into the result eq.(4.72), one finds α-values of about $51°$ for 15 mM salt, and a value of $64°$ for 5 mM, in line with the expectation that a decrease of salt concentration favours the opening of the structure.

The value of the entry-exit angle can be influenced by the histone tails if they bind to the DNA. This is a local effect, when the tails bind to the DNA attached to its own histone octamer.

There is, however, also the important possibility that the tails of different nucleosomes interact. This gives rise to a *tail-bridging effect* (H. SCHIESSEL, 2006), which is illustrated in Figure 4.5.

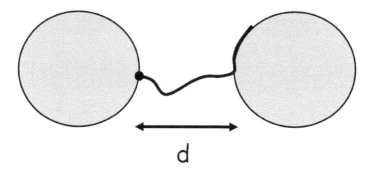

FIGURE II.4.5:   A tail bridging two spherical particles and the interaction potential between them. After H. SCHIESSEL, 2006.

Assuming that the nucleosomes can, in a zeroth-order approximation, as spherical particles of radius R, the interaction potential between them is given by a contribution of the repulsion between the two like-charges particles and the attraction caused by a bridge formed by a tail of contour length ℓ with the line density λ ,

$$U = U_{rep} + U_{bridge} \tag{4.74}$$

which is explicitly given by (ignoring precise numerical prefactors)

$$\frac{U(d)}{k_B T} = \frac{\ell_B Z^2}{(1 + R/\ell_D)^2} \frac{\exp - d/\ell_D}{d + 2R} \tag{4.75}$$
$$- \exp\left(-(d - \ell_D)\frac{\ell_B \ell_D Z\lambda}{R^2}\right) + \exp\left(-(\ell - \ell_D)\frac{\ell_B \ell_D Z\lambda}{R^2}\right) .$$

Based on this result one can study how the minimum of the potential changes as the line charge is varied. The minimum distance of the two particles, $d_{min} \sim \ln(C \cdot \lambda)$, where C is a constant given by the other parameters of the potential. The potential minimum becomes very shallow for decreasing λ, and finally vanishes.

Although these considerations give an idea about how electrostatic effects of the histone tails influence chromatin structure, there is still a large disparity between the understanding of any *specific* histone tail modification and the histone code hypothesis on the one side, and the physical compaction mechanism on the other. Given the highly unspecific nature of the electro-static interactions, and the apparent high specificity with which particular tail modifications can be placed and read, a satisfactory understanding of tran-scriptional regulation of chromatin seems still far out.

In particular, it is yet open whether the histone tail specifications, highly specific as they may be, really constitute a code. Recent experimental work by M. F. DION et al., 2005, on the acetylation of histone H4 provides evi-dence that only a very restricted number of these histone tail modifications are indeed 'coding' for specific regulatory processes. The conclusions support eearlier work by Q. REN and M. A. GOROVSKY, 2001, whose analysis of the acetylation patterns of a histone variant H2A.Z demonstrated the dominant role of rather non-specific electrostatic effects.

4.5 Bacterial chemotaxis: cooperativity once more

This final section in this Chapter on proteins considers a system which in-teracts cooperatively: a piece of the signal transduction pathway in bacteria. As it turns out, the modelling of the properties of this system needs both equilibrium and nonequilibrium methods. Although we limit ourselves here to the first approach, in some sense this section also prepares already for the models of dynamically interacting biomolecules of Part III of this book. But here we return, for the last time, to an example of the application of the Ising model in biology.

How do bacteria sense molecules, e.g., to find a food source? They make use of protein assemblies in their membranes, built up from transmembrane receptors which recognise molecules in the environment. The receptor assem-blies consist of well-known proteins and enzymes, a coupling protein CheW, and a histidine kinase, the enzyme CheA. In *E. coli*, these molecules are orga-nized at the cell poles in tight clusters in thousands of copies. How do these clusters organize themselves to sense molecules of their environment? As it turns out, it is the coupling of neighbouring receptors which enhances the *sensitivity* with which signals can be detected.

A physical model for the collective signalling of a receptor cluster has been built from the Ising model (Y. SHI and T. DUKE, 1999). We assume that the state of a receptor molecule i can be characterized by a variable V_i, which can experimentally be related to a conformational state of the molecule. The sig-nalling process occurs by the binding (docking) of a ligand which we describe by a variable B_i. We can then write

$$V_i = V_i(\{V_{j \neq i}\}, \{B_j\}). \tag{4.76}$$

where $i, j = 1, 2, ..., N$. Assuming for simplicity - supported by experimental evidence[9] - that V_i has only two values V_i^0 and V_i^1, we write for the state variable

$$V_i = \psi(\sum_j T_{ij} V_j + B_i - U_i) \tag{4.77}$$

where

$$\psi(x) = \begin{cases} V^1, & x > 0, \\ V^0, & x \leq 0. \end{cases} \tag{4.78}$$

In eq.(4.77), T_{ij} is the nearest-neighbour coupling between two receptors, and U_i some threshold value. This type of states is known in the literature on neural networks as a *McCulloch-Pitts threshold model*. We identify V^0 with the active conformation; a value $B_i > 0$ then favors the binding of an attractant.

With this construction, the receptor system is governed by the energy provided $T_{ij} = T_{ji}$ and $T_{ii} = 0$,

$$\mathcal{H} = -\sum_{\langle ij \rangle} T_{ij} V_i V_j - \sum_i B_i V_i + \sum_i U_i V_i. \tag{4.79}$$

This energy can be rewritten in Ising-spin notation (*Exercise!*) to yield

$$\mathcal{H} = -\sum_{\langle ij \rangle} J_{ij} S_i S_j - \sum_i H_i S_i \tag{4.80}$$

where additional constant terms have been ignored. So we have arrived at the Ising model - but hold on - isn't that a bit too crude a model?

Indeed, while the coupling between the receptors, mediated by J_{ij} seems very plausible, the ligand binding term merits some further interpretation. In fact, each receptor has two options: either a ligand is bound to it or it is not, and we decide that in the bound case, $H_i = H$, and otherwise $H_i = 0$.

---

[9]The two values of the state variable V correspond to different conformations of the protein, typically differently positioned helices.

That means we have to write down a *bimodal probability distribution* of the H_i:

$$p(H_i) = c\delta(H_i - H) + (1 - c)\delta(H_i), \tag{4.81}$$

where c is a measure of the concentration of the liganded receptors. The average over the H_i is then given by

$$\overline{H_i} = \int dH_i H_i p(H_i) = cH, \tag{4.82}$$

and the fluctuations of the random distribution of the H_i are

$$\Delta H_i = \sqrt{c(1 - c)}H. \tag{4.83}$$

We now solve this model in a mean-field theory approach and obtain

$$m = \overline{\tanh(\beta\nu Jm + \beta H_i)}, \tag{4.84}$$

where the overline indicates that the tanh still needs to be averaged over H_i. Explicitly, one finds that m obeys a transcendental equation,

$$m = \frac{2c}{1 + \exp[-2(\beta\nu Jm + \beta H)]} + \frac{2(1 - c)}{1 + \exp(-2\beta\nu Jm)} - 1. \tag{4.85}$$

Herein, ν is the number of nearest neighbors of the receptors, and $\beta = 1/k_B T$.

It is clear what these profiles will look like, since we are discussing an Ising model, after all: there will be two types of solutions generalizing the ferromagnetic and paramagnetic phases of the original Ising model. Of greater interest in the context of chemotaxis is the sensitivity to ligand docking, which we measure by the response of the order parameter m to a small concentration of ligands, hence

$$S \equiv \frac{1}{2}\partial_c m\,|_{c=0} = \frac{(1 + \exp(-2\beta H))^{-1} - 1/2}{1 - \beta\nu J}, \tag{4.86}$$

where $m(c = 0) = 0$ has been assumed.

Obviously, S can be made arbitrarily large for $\beta\nu J \to 1$, which is exactly the value at the phase transition. We see that it is indeed the large cooperative effect which makes the system so sensitive.

This model can be further extended in various ways, to include, e.g., a response to repellant molecules, to cover adaptation of the system to changes in environmental conditions, etc. We do not continue this here but refer to the literature, the paper by Shi and Duke, and others mentioned in the Additional

Notes.

Instead, we want to finish this Part with a remark on what message should be taken along to the next. In this Part we have seen a multifaceted picture involving aspects of *cooperative interactions* between biomolecules and their relationship to phase transitions; we have seen the role played by *disorder*, and we have met non-specific charge-dependent effects. These three determinants of the static interactions of biomolecules will now find counterparts in dynamic effects. We will see structure formation driven by nonlinear interactions, but also disordering and ordering effects due to fluctuations in the stochastic dynamics of molecules.

Additional Notes

Protein folding and docking. A modern standard reference on the protein folding problem is the book by A. FERSHT, 2003. The CI2-molecule is discussed in some detail.

There is an unconquerable amount of literature on protein folding. A recent listing of novel ideas is by D. J. BROCKWELL et al, 2000. The favourite approach within the statistical mechanics community is the development of as-simple-as-possible models to cover the large variety of observed structural features. A very recent prototypical example of this approach is the paper by T. X. HOANG et al., 2004. Note that the abstract begins "*We present a simple physical model...*".

Aspects of electrostatics for proteins are discussed in detail by B. HONIG and A. NICHOLS, 1995. Electrostatics for protein docking is reviewed in A. HEIFETZ et al., 2002. A general review of recent protein docking methods is by G. R. SMITH and M. J. E. STERNBERG, 2002.

Chromatin. A basic reference on the physics of chromatin is the review by H. SCHIESSEL, 2003. A basic introduction to the biological problems associated with chromatin is the book by B. M. TURNER, 2001.

We particularly mention three open problems of chromatin physics and biology:

Chromatin condensation. It is by now fairly clear that the mechanism underlying the condensation of chromatin is electrostatic in origin. The details of the process, and even more so, chromatin structure beyond the 10 nm fiber remain elusive so far. There has been some progress on the 30 nm fiber, in particular with sophisticated simulation methods J. LANGOWSKI and H. SCHIESSEL, 2004, J. SUN et al., 2005.

Nucleosome positioning. A problem of ongoing research on which much progress has been made recently is the position of the nucleosomes along the 10 nm fiber. Two aspects intervene here: the correlation of the positioning with the DNA sequence in contact with the nucleosome (M. S. COSGROVE et al., 2004, M. DLAKIĆ et al., 2004), and the dynamic repositioning/remodelling during transcriptional processes (H. SCHIESSEL et al., 2001, H. SCHIESSEL, 2006.

Histone code and dynamics of regulation. A problem of outstanding interest in modern biology is how the histone and histone tail modifications are written and read in conjunction with the transcriptional states on the DNA. So far, there is no quantitative model, but a number of competing ideas

(S. SCHREIBER and B. E. BERNSTEIN, 2002; A. BENECKE, 2003, M. S. COS-GROVE et al., 2004; W. FISCHLE et al., 2004, and G. LI and J. WIDOM, 2004.) Reviews of histone modifications are by B. M. TURNER, 2000, S. L. BERGER, 2002, and M. IIZUKA and M. M. Smith, 2003.

Bacterial chemotaxis. Bacterial chemotaxis is a problem which has been thoroughly investigated, both from a molecular as well as a 'systems' point of view. A lot of detail is known about the functional interactions between the receptors, see, e.g., the discussion by J. J. FALKE, 2002, and the recent work by V. SOURJIK and H. C. BERG, 2004. The latter paper presents a detailed discussion of molecular cooperativity in this system. A detailed kinetic model is by P. A. SPIRO, J. S. PARKINSON and H. G. ORTHMER, 1997. The system is also a prime example of the notion of 'robustness' in the context of biological systems, which refers to the systems' flexibility to respond to influences without requiring fine-tuning of system parameters. Seminal studies are N. BARKAI and S. LEIBLER, 1997, and U. ALON et al., 1999.

References

T. Agalioti, G. Chen, D. Thanos, *Deciphering the transcriptional Histone Acetylation Code for a human gene*, Cell **111**, 381-392 (2002)

U. Alon, M. G. Surette, N. Barkai and S. Leibler, *Robustness in bacterial chemotaxis*, Nature **397**, 168-171 (1999)

N. A. Baker, D. Sept, S. Joseph, M. J. Holst and J. A. McCammon, *Electrostatics of nanosystems: Application to microtubules and the ribosome*, Proc. Natl. Acad. Sci. USA **98**, 10037-10041 (2001)

A. Bannister and T. Kouzarides, *Reversing histone methylation*, Nature **436**, 1103-1106 (2005)

N. Barkai and S. Leibler, *Robustness in simple biochemical networks*, Nature **387**, 913-917 (1997)

A. Benecke, *Genomic plasticity and information processing by transcription coregulators*, ComPlexUs **1**, 65-76 (2003)

S. L. Berger, *Histone modifications in transcriptional regulation*, Curr. Op. Gen. Dev. **12**, 142-148 (2002)

D. J. Brockwell, D. A. Smith and S. E. Radford, *Protein folding mechanisms: new methods and emerging ideas*, Curr. Op. Struct. Biol. **10**, 16-25 (2000)

M. S. Cosgrove, J. D. Boeke and C. Wolberger, *Regulated nucleosome mobility and the histone code*, Nature Struct. Biol. **11**, 1037-1043 (2004)

B. Derrida, *Random-Energy Model: Limit of a Family of Disordered Models*, Phys. Rev. Lett. **45**, 79-82 (1980)

K. A. Dill and H. S. Chan, *From Levinthal to pathways to funnels*, Nat. Struct. Biol. **4**, 10-19 (1997)

M. F. Dion, S. J. Altschuler, L. F. Wu and O. J. Rando, *Genomic characterization reveals a simple histone H4 acetylation code*, Proc. Natl. Acad. Sci. USA **102**, 5501-5506 (2005)

M. Dlakić, D. W. Ussery and S. Brunak, *DNA Bendability and Nucleosome Positioning in Transcriptional Regulation*, in *DNA Conformation and Transcription*, Ch.14, T. Ohyama (ed.), Eurekah.com (2004)

J. J. Falke, *Cooperativity between bacterial chemotaxis receptors*, Proc. Natl. Acad. Sci. USA **99**, 6530-6532 (2002)

G. Felsenfeld and M. Groudine, *Controlling the double helix*, Nature **421**, 448-453 (2003)

A. Fersht, *Structure and Mechanics in Protein Science*, W. H. Freeman (2003)

W. Fischle, Y. Wang and C. D. Allis, *Binary switches and modification cassettes in histone biology and beyond*, Nature **425**, 475-479 (2003)

A. Heifetz, E. Katchalski-Katzir and M. Eisenstein, *Electrostatics in protein-protein docking*, Protein Science **11**, 571-587 (2002)

A. Hildebrandt, R. Blossey, S. Rjasanow, O. Kohlbacher, H.-P. Lenhof, *Novel formulation of nonlocal eletrostatics*, Phys. Rev. Lett. **93**, 108104 (2004)

A. Hildebrandt, *Biomolecules in a structured solvent*, Rhombos (2005)

T. X. Hoang, A. Trovato, F. Seno, J. R. Banavar and A. Maritan, *Geometry and symmetry presculpt the free-energy landscape of proteins*, Proc. Natl. Acad. Sci. USA **101**, 7960-7964 (2004)

B. Honig and A. Nichols, *Classical Electrostatics in Biology and Chemistry*, Science **268**, 1144-1149 (1995)

M. Iizuka and M. M. Smith, *Functional consequences of chromatin modifications*, Curr. Op. Gen. Dev. **13**, 154-160 (2003)

J. D. Jackson, *Classical Electrodynamics* 3rd ed., Wiley (1999)

R. M. Jackson and M. J. E. Sternberg, *A Continuum Model for Protein-Protein Interactions: Application to the Docking Problem*, J. Mol. Biol. **250**, 258-275 (1995)

J. Janin and S.J. Wodak, *Computer analysis of protein-protein interactions*, J. Mol. Biol. **124**, 323-342 (1978)

E. Katchalski-Katzir, I. Shariv, M. Eisenstein, A. A. Friesem, C. Afaflo, I. A. Vakser, *Molecular Surface Recognition: Determination of Geometric Fit Between Proteins and Their Ligands by Correlation Techniques*, Proc. Natl. Acad. Sci. USA **89**, 2195-2199 (1992)

A. A. Kornyshev, A. I. Rubinstein and M. A. Vorotyntsev, *Model nonlocal electrostatics: I*, J. Phys. C: Solid State Phys. **11**, 3307-3322 (1978)

A. A. Kornyshev and M. A. Vorotyntsev, *Model nonlocal electrostatics: III. Cylindrical interface*, J. Phys. C: Solid State Phys. **12**, 4939-4946 (1979)

J. Langowski and H. Schiessel, *Theory and computational modelling of the 30 nm chromatin fiber*, New Comprehensive Biochemistry **39**, 397-420 (2004)

C. Levinthal, *How to fold graciously*, in *Mössbauer Spectroscopy in Biological Systems*, P. Debrunner and J. C. M.Tsibris (eds.), 22-24 (1968)

G. Li and J. Widom, *Nucleosomes facilitate their own invasion*, Nat. Struct. Mol. Biol. **11**, 763-769 (2004)

K. Luger, A. W. Mäder, R. K. Richmond, D. F. Sargent and T. J. Richmond, Nature **389**, 251-260 (1997)

G. S. Manning, *Limiting Laws and Counterion Condensation in Polyelectrolyte Solutions I.*, J. Chem. Phys. **51**, 924-933 (1969)

C. Merlo, K. A. Dill and T. R. Weikl, Φ-*values in protein-folding kinetics have energetic and structural components*, Proc. Natl. Acad. Sci. USA **102**, 10171-10175 (2005)

J. N. Onuchic, Z. Luthey-Schulten and P. G. Wolynes, *Theory of Protein Folding: The Energy Landscape Perspective*, Annu. Rev. Phys. Chem. **48**, 545-600 (1997)

Q. Ren and M. A. Gorovsky, *Histone H2A.Z Acetylation Modulates an Essential Charge Patch*, Molecular Cell **7**, 1329-1335 (2001)

H. Schiessel, J. Widom, R. F. Bruinsma and W. M. Gelbart, *Polymer Reptation and Nucleosome Repositioning*, Phys. Rev. Lett. **86**, 4414-4417 (2001)

H. Schiessel, *How short-ranged electrostatics controls the chromatin structure on much larger scales*, Europhys. Lett. **58**, 140-146 (2002)

H. Schiessel, *The physics of chromatin*, J. Phys. Cond. Mat. **15**, R699-R774 (2003)

H. Schiessel, *The nucleosome: A transparent, slippery, sticky and yet stable DNA-protein complex*, Eur. Phys. J. E **19**, 251-262 (2006)

S. Schreiber and B. E. Bradstein, *Signaling Network Model of Chromatin*, Cell **111**, 771-778 (2002)

Y. Shi and T. Duke, *Cooperative model of bacterial sensing*, Phys. Rev. E **58**, 6399-6406 (1998)

T. Simonson, *Macromolecular electrostatics: continuum models and their growing pains*, Curr. Op. Struct. Biol. **11**, 243-252 (2001)

G. R. Smith and M. J. E. Sternberg, *Prediction of protein-protein interactions by docking methods*, Curr. Op. Struct. Biol. **12**, 28-35 (2002)

V. Sourjik and H. C. Berg, *Functional interactions between receptors in bacterial chemotaxis*, Nature **428**, 437-441 (2004)

P. A. Spiro, J. S. Parkinson and H. G. Othmer, *A model of excitation and adaptation in bacterial chemotaxis*, Proc. Natl. Acad. Sci. USA **94**, 7263-7268 (1997)

B. D. Strahl and C. D. Allis, *The language of covalent histone modifications*, Nature **403**, 41-45 (2000)

J. Sun, Q. Zhang and T. Schlick, *Electrostatic mechanism of nucleosomal array folding revealed by computer simulation*, Proc. Natl. Acad. Sci. USA **102**, 8180-8185 (2005)

B. M. Turner, *Histone Acetylation. A global regulator of chromatin function*, in C. Nicolini (ed.), Genome Structure and Function, Kluwer Academic Publishers, 155-171 (1997)

B. M. Turner, *Chromatin and gene regulation*, Blackwell Science (2001)

B. M. Turner, *Cellular Memory and the Histone Code*, Cell **111**, 285-291 (2002)

M. A. Vorotyntsev, *Model nonlocal electrostatics: II, Spherical interface*, J. Phys. C: Solid State Phys. **11**, 3323-3331 (1978)

P. Wolynes, *Landscapes, Funnels, Glasses and Folding: From Metaphor to Software*, Proc. Americ. Philosoph. Soc. **145**, 555-563 (2001)

Part III

Networks

Chapter 1

Network Dynamics I: Deterministic

How come all my body parts so nicely fit together?
All my organs doing their jobs, no help from me!

Crash Test Dummies, How does a duck know? (1993)

The detailed knowledge of the properties of biomolecules and their interactions, as we have described them in Part II of this book, is a necessary but not a sufficient step to understand biological systems. The next step is to understand how the different molecular components *interact in concert* and how they build up the hierarchy observed in biological systems: cellular compartments, cells, organs and organisms.

In most of what follows, we will be interested here in *genetic networks* of some sort. Under this term we understand biological reaction schemes on the basis of the central dogma, DNA makes RNA makes protein. We will encounter the following situations:

- A protein attaches itself to DNA at a binding site; such molecules are *transcription factors* which help control gene expression;

- A protein-RNA complex, the RNA polymerase, reads out the genes after fixing itself at a *promoter site*; the readout is a messenger RNA;

- the mRNA transcript is read at a ribosome and serves as a blueprint for a protein;

- proteins compete with each other; this may happen by their direct interaction or blocking or activating the corresponding genes.

This basic readout mechanism underlies the complexity of all gene networks. The examples we discuss have been selected to allow for - mostly - analytical calculations to illustrate several aspects arising in *systems biology*.

We will build this up in a step-wise fashion and progress through a description of a simple regulatory circuit, modelled by deterministic differential equations and move on to the emergence of spatial structures. This material is covered in Chapter 1. Chapter 2 is concerned with the effects of fluctuations on the regulatory components. Finally, in Chapter 3, we will take a more global view and have a look at the overall structure of biological networks.[1]

1.1   Deterministic dynamics: λ-repressor expression

In this section we want to discuss an example of a biological network - a very small one - and to formulate a simple model for a particular aspect of its dynamics. The example is the λ-phage.

The λ-phage is a bacterial virus for which the molecular basis of its life cycle is very well understood (M. PTASHNE, 2004). Consequently, it is often used as a prototypical example for theoretical studies. Here, we use it as a 'simple' complex system to illustrate how to describe biological networks within a deterministic dynamics setting. But first, we need to learn some basics of the life cycle of the virus; many more details can be found in Ptashne's book.

Life cycle of phage λ. Phage λ is a bacterial virus which infects *E. coli*. The virus inserts its DNA into the bacterial cell. Here, the DNA has two options. The first is its insertion into the host genome, and it then 'stays on board' as a 'blind passenger', being passed on from one generation to the next upon cell division. This phase is called the *lysogenic phase*. But if, e.g., the bacterium is endangered and responds to this external stress, the virus can 'abandon ship'. Its genome is excised out of the host genome and begins the production of new viruses which then leave the host (which dies). The phase in which the virus replicates is called the *lytic phase*.

The decision making process between the lysogenic or lytic phase is made at a molecular switch. The switch is illustrated schematically in Figure 1.1. It consists of an *operon* built from three operator binding sites, at which a dimer of the λ-repressor molecule can attach itself. The λ-repressor is a transcription factor, i.e., its presence on DNA serves as a flag for RNA polymerase to attach itself and start transcription of an adjacent gene. Indeed, the three

---

[1]The insider immediately recognizes this progression as typical for the statistical physics approach: for few components, detailed models can be built - while things again simplify when a large number of components can be studied. Admittedly, the real difficulty lies in the 'middle thing'.

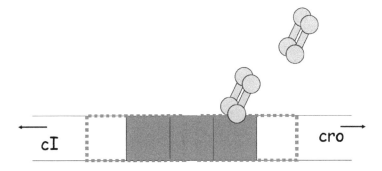

FIGURE 1.1: The λ-phage operon. Operator sites are shown as gray-shaded areas, with OR3-OR2-OR1 from left to right. Promoters are indicated by dashed frames; gene transcription directions are indicated by arrows. Two λ-repressor dimers are shown upon attachment and approach. (After M. PTASHNE, 2004.)

operator sites are at the same time overlapping with two neighbouring promoter regions, which are the fixation platforms for RNA polymerase, and they directly neighbour two adjacent genes, *cI*, the gene for the λ-repressor molecule, and *cro*, the gene for a second transcription factor also involved in the switch. Although operator and promoter sites are overlapping - and part of their function depends on this fact - their biological role is different.

The switch now functions as follows. The attachment of a CI-dimer (the other name of the dimer form of the λ-repressor) at operator site OR1 enhances the fixation probability of a second dimer to fix at OR2. The fixation of CI at OR2, in turn, increases the probability of fixation of RNA polymerase at the *cI*-promoter, upon which transcription of *cI* occurs. The action of *cI* is thus directly implied in its description, hence the gene is *autoregulatory*. The continual transcription of *cI* ensures that the system stays in the lysogenic phase.

If, on the other hand, the system needs to switch to the lytic phase, a repressor dimer can fix at OR3, which due to its overlap with the *cI*-promoter blocks the access for RNA-polymerase. Now, *Cro*-proteins intervene, and they attach themselves to the same OR-sites as CI, but with inverted affinities. A *Cro*-protein present at OR2 enhances the fixation probability for a polymerase at the *cro*-promoter, and transcription of *cro* can start, initiating the lytic pathway.

Let's now build some of these mechanisms into a mathematical model.

Hasty model I: basics. In this subsection we discuss a simple model for repressor expression proposed by J. HASTY et al., 2000. The main actor in this model is the repressor molecule itself; we denote it by X. The advantage of the Hasty model is that it allows to quickly gain a qualitative insight, but it can be made fully quantitative as well. Here, we are mainly interested in the qualitative aspects, though.

A first simplification in the Hasty model is the assumption that one can neglect the operator binding site OR1. This is justified if one is only interested in the repressor kinetics: it means we do not want to model the full switch, but just the autoregulation mechanism of the λ-repressor.

In this case, the system can be described by six reactions. Four of them will be assumed as being *fast* when compared to the others; these are the association and dissociation of the transcription factors from their binding sites. They will be considered as equilibrium reactions, and are given by

- Repressor dimerization, $2X \leftrightarrow X_2$ with equilibrium constant K_1;

- Repressor dimer binding to the DNA promoter site OR2 on D, $D+X_2 \leftrightarrow DX_2$, with equilibrium constant K_2;

- Repressor dimer binding to the DNA promoter site OR3 on D, $D+X_2 \leftrightarrow DX_2^*$, with equilibrium constant K_3;

- Repressor dimer binding to the DNA promoter sites OR2 and OR3 on D, $DX_2 + X_2 \leftrightarrow DX_2X_2$, with equilibrium constant K_4.

The slow reactions in the system are transcription and degradation. Transcription is described by the irreversible reaction scheme

$$DX_2 + P \rightarrow_{k_t} DX_2 + P + nX \tag{1.1}$$

in which P is the DNA polymerase, and n is the number of repressor proteins per RNA transcript. The degradation reaction reads

$$X \rightarrow_{k_2} \phi\,. \tag{1.2}$$

We now define concentrations for all variables,

$$x \equiv [X],\ y \equiv [X_2],\ d \equiv [D],\ u \equiv [DX_2],\ v \equiv [DX_2^*],\ z \equiv [DX_2X_2]. \tag{1.3}$$

Considering the fast reactions as equilibrium reactions allows to write them as simple algebraic expressions,

$$y = K_1 x^2 \tag{1.4}$$
$$u = K_2 dy = K_1 K_2 dx^2 \tag{1.5}$$
$$v = \sigma_1 K_2 dy = \sigma_1 K_1 K_2 dx^2 \tag{1.6}$$
$$z = \sigma_2 K_2 uy = \sigma_2 (K_1 K_2)^2 dx^4 . \tag{1.7}$$

It now remains to write down a rate equation for the repressor, which is given by

$$\dot{x} = \tilde{\alpha} u - k_d x + r . \tag{1.8}$$

Here, $\tilde{\alpha} = n k_t p_0$ is a constant containing k_t and the concentration of polymerase, p_0, which is assumed fixed. The term $-k_d x$ describes degradation of x while the last term, r, constitutes a basal transcription rate. In order to close the system of equations, a conservation law needs to be invoked. In fact, the total concentration of DNA promoter sites, d_T, is fixed.[2] Thus

$$d_T = d + u + v + z = d(1 + (1 + \sigma_1)K_1 K_2 x^2 + \sigma_2 K_1^2 K_2^2 x^4) \tag{1.9}$$

which leads to

$$d = \frac{d_T}{1 + (1 + \sigma_1)K_1 K_2 x^2 + \sigma_2 K_1^2 K_2^2 x^4} . \tag{1.10}$$

This use of expression eq.(1.10) allows us to rewrite eq.(1.8) in a succinct form containing only x as a variable,

$$\dot{x} = \frac{\alpha x^2}{1 + 2x^2 + 5x^4} - \gamma x + 1 , \tag{1.11}$$

where repressor concentration and time have been rescaled to dimensionless quantities, and the values $\sigma_1 \approx 1$, $\sigma_2 \approx 5$, obtained from experimental estimates, have been used. The remaining parameters α and γ then constitute ratios of transcription rate and degradation rate relative to the basal transcription rate.

The mathematical discussion of this equation is extremely simplified by its one-dimensional character. In fact, we can rewrite it in the form

$$\dot{x} = -\partial_x V(x) \tag{1.12}$$

---

[2]This is true *in vivo*, for small concentrations, but even more so *in vitro*, for large concentrations.

where $V(x)$ is the integral of the right hand side of eq.(1.12). $V(x)$ is now seen to serve as a 'potential energy' landscape, shown in Figure 1.2, a function which has two minima of different depth. Whichever is the lower of the two determines the steady state of the repressor system, i.e., either a state with a low concentration, or with a high concentration of repressor molecules.[3] Figure 1.3 is the main result of this subsection; it shows that repressor expression is a bistable system arising from the two competing mechanisms of repressor production and degradation.

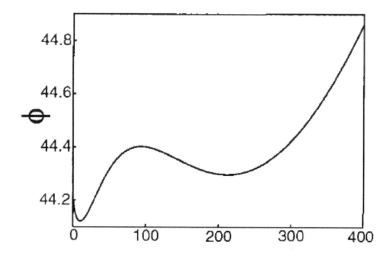

FIGURE 1.2:   The effective potential of the λ-repressor. From J. HASTY et al., Copyright (2000) National Academy of Sciences, U.S.A.

Hasty model II: looping included. We now include *looping* into the model. What this means is indicated in Figure 1.4. In the λ-phage a second, or left, operator region (OL) is found along the genome. This region, to which the corresponding genes are lacking, has a peculiar regulatory function (B. RÉVET et al., 1999). Repressors that fix themselves at the OL-sites, are present to interact with their cousins sitting at OR-sites by forming a repressor octamer. Octamerization of repressors in solution is a very rare event, but the presence of the molecules on the DNA increases the formation rate enormously, since looping is a very effective mechanism of bringing different regions of DNA into contact (H. MERLITZ et al., 1998). It is therefore also a very common regulatory mechanism in eukaryotes.

---

[3]This argument, as nice as it is, works indeed in general only for one variable. For more than one variable, a potential will only exist in very special cases.

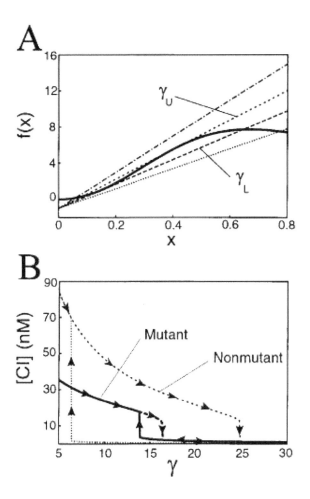

FIGURE 1.3:   Bistability in phage λ. Graph A shows the graphical solution of the stationary points of eq.(1.12), i.e., $\dot{x} = 0$. In the figure, the solid curve is the first term of the rhs of eq.(1.12), while the broken lines are given by the function $f(x) = \gamma x - 1$. The intersections of the graphs correspond to the extrema of $V(x)$, and can, e.g., be varied by changing γ. Graph B shows a hysteresis diagram of repressor concentration vs. γ. The size of the hysteresis loop is a simple criterion for the barrier between both states: the bigger the loop, the higher the barrier. The non-mutant case refers to the situation in which all operator sites are present, while the mutant case corresponds to the simplified model we discussed in the text. Qualitatively, both cases are similar, but the quantitiative difference is significant. From J. HASTY et al., Copyright (2000) National Academy of Sciences, U.S.A.

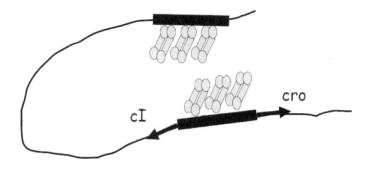

FIGURE 1.4: Looking in phage λ. After M. PTASHNE, 2004.

The Hasty model can be extended to account for DNA looping; one needs two additional assumptions. Again we simplify the structure of the left operator as we did for the right:

- the second operator unit equals the first: OR = OL; this in particular refers to all rate constants. OL is placed on a second DNA site, d_L.

- there is an additional kinetic process described by the complex formation between a doubly bound repressor at OR and the same at OL. This process has a rate K_5, and is assumed to be in equilibrium (fast compared to repressor production).

The presence of this additional mechanism modifies the nonlinearity in eq. (1.12), which we now call $f_\ell(x)$, in the case looping. The change is due to the different possibilities of repressor fixation at either OR or at OL (d_R or d_L). Assuming the symmetric case with $d_R = d_L = d_\ell/2$ we find the conservation law

$$d_T = d_\ell(1 + (1 + \sigma_1)x^2 + \sigma_2 x^4) + \frac{\delta d_\ell^2}{4} x^8 \tag{1.13}$$

which replaces eq.(1.9), and where the factor δ derives from the looping process, i.e., the kinetic rate K_5. When the equation is solved for d_ℓ this yields, with $\sigma_1 = 2$ and $\sigma_2 = 5$,

$$d_\ell = 2 \left[\frac{1 + 2x^2 + 5x^4}{\delta x^8} \right] \left[\left[1 + \frac{d_T \delta x^8}{(1 + 2x^2 + 5x^4)^2} \right]^{1/2} - 1 \right]. \tag{1.14}$$

Although expression (1.14) looks rather different from (1.10), there is in fact only a minor difference between them. This fact becomes obvious by expanding the square root in eq.(1.14) for small x, which obviously reduces (1.14) to the eq.(1.10), while for large x the curves have the same asymptotics but a different amplitude.

This insensitivity to the presence of the nonlinear coupling is astonishing, but it allows to understand the role played by the looping-mediated coupling for the repressor dynamics. If one compares the stationary points of the repressor system with and without looping one sees that the effect of looping is simply to reduce the bistable region in which repressed and non-repressed transcriptional states coexist, see Figure 1.3. Given that the full operator region indeed has a significantly larger bistable region due to the presence of the operator site OR1 which we neglected in the calculation, the looping mechanism can be understood as an effective means to dynamically modulate the bistability of the repressor autoregulation.

As a further consequence, the looping mechanism also affects the fluctuations, which we have not discussed at all. Since, with looping allowed for, the system can switch between the two states within a much narrower concentration range, the frequency of switching will consequently increase, while the amplitude diminishes. This is in accord with recent conclusions based on stochastic simulations (J. M. G. VILAR and S. LEIBLER, 2005).

It is worth noting that the simple model presented here neglects all spatial structure of the looping mechanism; the mechanism appears identical to that of an octameric complex binding at OR. But it is this process which is highly unlikely to occur since it has a very small equilibrium constant. By contrast, bringing dimer complexes at sufficiently distant sites into contact via looping occurs with a higher probability. The binding of repressor dimers at the OL region thus constitutes an effective mechanism to hold the molecules 'in stock' by placing them on the DNA, rather than letting them diffuse through solution.

1.2 The Turing insight

However, a partial differential equation will be obtained which is thought to give a good approximation to the behaviour of certain kinds of chemical system. The differential equation has a number of parameters and it is necessary to find values for these parameters which will make the differential equation behave properly.

Alan Turing, The development of the daisy (1952)

In the 1950s Alan Turing hypothesized that the formation of biological structures may be driven by chemical instabilities. He proposed that the patterning results from an instability involving the reaction and diffusion of particular molecules, the *morphogens*. The idea that structure formation is based on the spatial distribution of chemical compounds and their interactions has proved to be a very profound one; meanwhile, several different variations of this scheme exist. We will go through the most important ones.

Morphogen gradients. The simplest way to generate a pattern is to assume that a morphogen is present at some localized source, and then diffuses into the surrounding tissue. The simplest chemical reaction that can happen is the decay of the morphogen due to degradation. Such a system is described by the differential equation

$$\partial_t \phi = D\partial_x^2 \phi - \mu\phi^\alpha + \eta\delta(x)\,, \tag{1.15}$$

where ϕ is the concentration of the morphogen, D is its diffusion coefficient in the tissue, and μ the degradation coefficient. The delta-function term describes the localized source of the morphogen. Note that we have restricted ourselves to a one-dimensional case.

The parameter α allows to tune the degradation process: for $\alpha = 1$ one has the 'simple' degradation process while the value $\alpha = 2$ corresponds to an enhanced autodegradation. This case has recently been studied by (J. L. ENGLAND and J. CARDY, 2005). Eq.(1.15) then has a steady-state solution $\phi_0(x)$ given by

$$\phi_0(x) = \frac{6D}{\mu}\frac{1}{(x+\epsilon)^2} \tag{1.16}$$

where $\epsilon \equiv (12D^2/\mu\eta)^{1/3}$. This solution corresponds to a gradient profile which decays algebraically outwards from the source.

A morphogen gradient can serve as a very simple means to structure s-pace. Any mechanism capable to detect *absolute* concentrations, say a value ϕ^*, naturally devides space into regimes $\phi > \phi^*$ and $\phi < \phi^*$. Although this mechanism seems astonishingly simple, it is actually realised in many systems (see, e.g., J. B. GURDON and P. Y. BOURILLOT, 2001).

Activators and inhibitors. More complex structuring effects can arise by considering not only one substance which decays, but also the interaction of two substances, one of which we call 'activator' a, while the second is called an 'inhibitor', h. A possible set of *reaction-diffusion equations* for such a system is given by

$$\partial_t a = D_a \partial_x^2 a - \mu a + \frac{\varrho a^2}{h} + \varrho_a , \tag{1.17}$$

$$\partial_t h = D_h \partial_x^2 h - \nu h + \varrho a^2 + \varrho_h , \tag{1.18}$$

where the terms $\propto \varrho a^2$ describe the auto- and crosscatalysis of the activator - as opposed to an autodegradation we had in eq.(1.15) - while the term $\propto 1/h$ covers the action of the inhibitor; the remaining terms are degradation and source terms, as one verifies upon comparison with eq.(1.15).

Equations (1.17), (1.18) are examples of a large class of models which were shown to generate patterns by a combination of *short-range activation* due to the catalytic nonlinearity $\propto a^2$ and a *long-range inhibition*. Several examples with different types of nonlinearities are discussed in the seminal paper by (A. GIERER and H. MEINHARDT, 1972). We will come back on the choice of nonlinearities later on.

We now turn to the technical discussion of the linear instability in the reaction-diffusion systems.

The Turing model. Turing was the first to see that *instabilities* in reaction-diffusion systems can lead to a pattern formation process. Let us discuss the appearance of an instability in the reaction-diffusion equations.

Suppose there are two chemical substances of concentrations ϕ_1 and ϕ_2, each of which is allowed to react according to a chemical reaction scheme (see Part I, Chapter 2). As before, the substances may diffuse. We now write down a general system for the reaction-diffusion equations

$$\partial_t \phi_1 = f(\phi_1, \phi_2) + D_1 \Delta \phi_1 \tag{1.19}$$
$$\partial_t \phi_2 = g(\phi_1, \phi_2) + D_2 \Delta \phi_2$$

wherein f and g are nonlinear functions of ϕ_1 and ϕ_2, $D_{1,2}$ are diffusion constants, and $\Delta \equiv (\partial_x^2 + \partial_y^2 + \partial_z^2)$ is the Laplace operator in cartesian coordinates in three dimensions, generalizing what we wrote down before in eqs.(1.17), (1.18).

Suppose now that this system of equations has a *stationary point*, $\partial_t \phi_1 = \partial_t \phi_2 = 0$ at $(\phi_1, \phi_2) = (0,0)$. A linearization of eqs.(1.19) around this state in the form $\phi = \phi_0 + \delta\phi$ then leads to a linear system in the fluctuations $\delta\phi$; we drop the δ in the following to simplify notation:

$$\frac{d}{dt}\begin{pmatrix} \phi_1 \\ \phi_2 \end{pmatrix} = \begin{pmatrix} a_{11} & a_{12} \\ a_{21} & a_{22} \end{pmatrix} \cdot \begin{pmatrix} \phi_1 \\ \phi_2 \end{pmatrix} + \begin{pmatrix} D_1 \Delta \phi_1 \\ D_2 \Delta \phi_2 \end{pmatrix}. \tag{1.20}$$

The coefficients a_{ij} will in general depend on the value of ϕ_0. Now suppose that the system is confined to a cubic volume. Assuming that there is no flux of ϕ_1 or ϕ_2 entering the cube of side length ℓ, a *Neumann boundary condition* holds, i.e., the first spatial derivative of $\phi_{1,2}$ with respect to the wall normal vanishes.

The solutions of eqs.(1.20) can then be written in a general form as

$$\phi_1(\mathbf{x}, t) = \sum_{n_1, n_2, n_3 \geq 0} c_{n_1, \ldots, n_k}(t) \cos\left(\frac{2\pi}{\ell} x_1\right) \cdots \cos\left(\frac{2\pi}{\ell} x_3\right)$$

$$\phi_2(\mathbf{x}, t) = \sum_{n_1, n_2, n_3 \geq 0} d_{n_1, \ldots, n_k}(t) \cos\left(\frac{2\pi}{\ell} x_1\right) \cdots \cos\left(\frac{2\pi}{\ell} x_3\right)$$

$$\tag{1.21}$$

where $c_{..}, d_{..}$ are the *Fourier coefficients* of the solution to eqs.(1.20). The terms in the sum are the eigenmode solutions, indexed by the tuple set $N \equiv (n_1, n_2, n_3)$. The introduction of these expressions into eqs.(1.20) then leads to an infinite system of ordinary differential equations for the Fourier coefficients, given by

$$\frac{d}{dt}\begin{pmatrix} c_N \\ d_N \end{pmatrix} = \begin{pmatrix} a_{11} - (\frac{2\pi}{\ell})^2 D_1 |N|^2 & a_{12} \\ a_{21} & a_{22} - (\frac{2\pi}{\ell})^2 D_1 |N|^2 \end{pmatrix} \cdot \begin{pmatrix} c_N \\ d_N \end{pmatrix} \tag{1.22}$$

where $|N|^2 = n_1^2 + n_2^2 + n_3^2$.

We denote, in what follows, the (2×2)-matrix in eq.(1.22) by M_N. For each non-negative 3-tuple N the stability of the eigenmodes is determined by the eigenvalues of M_N, $\lambda_N^{\pm}$. These can be expressed in terms of the trace and the determinant of the matrix M_N,

$$\lambda_N^{\pm} = \frac{1}{2}\left(TrM_N \pm \sqrt{(TrM_N)^2 - 4DetM_N}\right). \tag{1.23}$$

For every $N \geq (0,0,0)$ the real and imaginary parts of eq. (1.23) are bounded from above; the value

$$\Lambda \equiv \max \, Re \, \lambda_N^{\pm} \qquad (1.24)$$

thus is the upper bound of the spectral abscissae of the set of matrices $\{M_N\}$ (in the positive quadrant).

We see already by inspection of the entries of the matrix M_N that the diffusion terms may indeed destabilize a stable state of the homogeneous system, which may lead to a *symmetry breaking* of the global behaviour of the solutions to the nonlinear equations. As it turns out, for the generation of patterns, the two diffusion constants must significantly differ from each other, which may not be evident in many biological situations: why should two proteins that, say, diffuse in the cytoplasm have significantly different diffusion constants? One way out, as we will see below, is that one diffusion process is confined to the cell membrane, while the other occurs in the cytoplasm.

Returning to the eigenvalue problem of the Turing system, there are two types of eigenvalues that indicate the linear instability of the homogeneous state: in case of a *Turing instability*, Λ is positive and is one of the eigenvalues of M_N; if there is an oscillatory instability, there exists a mode with $Re \, \lambda_N^{\pm} = \Lambda > 0$, but for which $Im \, \lambda_N^{\pm} \neq 0$.

The occurrence of a Turing instability is, however, not generally sufficient to generate a *Turing pattern*, i.e., a time-independent spatial structure. We do not follow this trace here, since it is not necessarily a stationary pattern that is of interest to us in the description of spatial structures in biological systems. Also, patterning may arise in systems in which the conditions built into, e.g., the Gierer-Meinhardt model, like the production of an activator, need not be the case. We will now see a particular example, the Min system, to which this applies.

1.3   The Min system

In this section, we study an example of a pattern forming processes: the cell division mechanism in *E. coli*. *E. coli* is a rod shaped bacterium of 2-6 μm length and 1-1.5 μm diameter. It divides roughly every hour: after replication of the DNA, a cell splits in half to form two daughter cells. The question is: how does the cell 'know' where its middle is?

The mechanism for cell division is based on the action of three proteins, the *Min* family: MinC, MinD and MinE. These proteins oscillate from one end to the other within the cytoplasm, and, furthermore, they move between the cytoplasm and the cytoplasmic membrane. Their concentrations determine the site for the cell division to occur.

The cell division itself is brought about by the formation of a contractile polymeric ring, called Z-ring, at the site of division. This ring is made from a protein named FtsZ, which is a homologue of tubulin which plays a similar role in higher organisms. FtsZ forms just underneath the cytoplasmic membrane, but how it generates the force needed for cell devision is still unknown. The Min-system serves to confine the location of this ring to midcell. If this mechanism is perturbed, ring formation can occur in the vicinity of the cell poles, and can lead to the formation of minicells devoid of DNA.

The action of the Min-system, as far as it is now known from molecular biology studies based on deletion mutants (i.e., bacteria in which corresponding genes have been knocked out), is as follows:

- MinC localized in the cytoplasmic membrane locally inhibits assembly of the contractile Z-ring. It remains in the cytoplasm and inactive in the absence of MinD;

- MinD binds MinC and recruits it to the cytoplasmic membrane;

- MinE drives MinD away from the membrane; the fact that it is driven mostly away from the bacterial midplane and hence allows for ring formation is a result of the protein dynamics.

Without MinE, the MinC/MinD couple would inhibit Z-ring formation everywhere, blocking cell division. In this case, long filamentous cells are observed. Without MinC, Z-ring formation cannot be inhibited anywhere, leading to inviable cells. Without MinD, neither MinC nor MinE are recruited to the cytoplasmic membrane and hence have less effect.

The diffusion dynamics of the Min-system has been visualized by fluorescence techniques. Hereby, the Min proteins are modified to fusion proteins with GFP (*green fluorescent protein*). This is illustrated in Figure 1.5. The MinC/Min D accumulate at a polar end of the bacterium on the cytoplasmic membrane. MinE forms a band at midcell which sweeps to the polar end and ejects the MinC/MinD into the cytoplasm. The ejected proteins then rebind at the other end of the bacterium. The MinE band, by contrast, dissociates at the pole and reforms at the center, and the whole process repeats towards the opposite cell pole. The observed oscillation period lasts 1-2 min, well below the cell division cycle length. The net effect of the dynamics is to minimize the concentration of MinC/MinD at midcell, and maximizing MinE concentration there, see Figure 1.6. The result shown there is obtained from a model

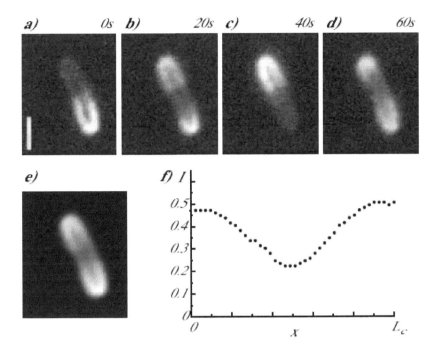

FIGURE 1.5: MinD oscillations in *E. coli*. The pictures a)-d) show fluorescence images of MinD-GFP at subsequent time points separated by 20 s; the scale bar in a) is 1 μm. Picture e) shows a time-average of frames separated by 1s during one full oscillation cycle. Finally, Picture f) shows the fluorescence intensity I from a line scan of the time under e). The total cell length is $L_c = 2.3\,\mu$m. (Reprinted with permission from G. MEACCHI and K. KRUSE, 2005.)

developed by M. HOWARD ET AL., 2001, to which we will now turn.

The quantitative model for this dynamics based on the reaction-diffusion equation proposed by M. HOWARD et al., 2001, is based on four coupled equations for the densities of MinD and MinE in the cytoplasm and the cytoplasmic membrane. Since the MinC dynamics follows that of MinD, it does not need to be considered explicitly. The equations read (here, densities are ϱ_i with $i = e, E$ and $i = d, D$ for the membrane (small) and cytoplasm (capital)):

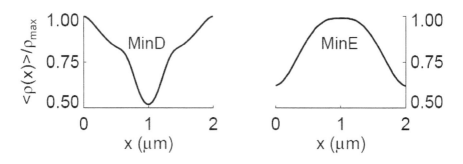

FIGURE 1.6: Concentration profiles of the Min proteins in *E. coli*. (M. HOWARD et al.; Copyright (2001) American Physical Society)

$$\partial_t \varrho_D = D_D \partial_x^2 \varrho_D - \frac{\sigma_1 \varrho_D}{1 + \sigma_1' \varrho_e} + \sigma_2 \varrho_e \varrho_d \qquad (1.25)$$

$$\partial_t \varrho_d = \frac{\sigma_1 \varrho_D}{1 + \sigma_1' \varrho_e} - \sigma_2 \varrho_e \varrho_d \qquad (1.26)$$

$$\partial_t \varrho_E = D_E \partial_x^2 \varrho_E - \sigma_3 \varrho_E \varrho_D + \frac{\sigma_4 \varrho_E}{1 + \sigma_4' \varrho_D} \qquad (1.27)$$

$$\partial_t \varrho_e = \sigma_3 \varrho_E \varrho_D - \frac{\sigma_4 \varrho_e}{1 + \sigma_4' \varrho_D} \qquad (1.28)$$

The first reaction terms in the equations for MinD describe the spontaneous association of MinD to the cytoplasmic membrane. Cytoplasmic MinD recruits cytoplasmic MinE to the membrane via the reaction term in the equation for MinE. Once there, MinE drives MinD into the cytoplasm, given by the second reaction term in the equations for MinD. The last term in the equations for MinE corresponds to the spontaneous dissociation from the membrane. All these terms have been modelled by introducing a parameter set $\{\sigma\}$, with values given below.

This dynamics conserves protein number; there are no source or drain terms. The model exhibits a linear Turing-like *Hopf-instability*. Note that the mechanism of pattern formation in this system differs from the one proposed by Turing, and Meinhardt and Gierer, since in this system there is no production of activators or an external influx of particles. The origin of the instability is the disparity between the diffusion rates between the cytoplasm and the cytoplasmic membrane; this has been idealized by setting the diffusion rates in the membrane to zero.

Task. Analyze the linear stability of the system, testing for solutions of the

TABLE 1.1: Min system parameters

D_D ($\mu m^2/s$)	D_E ($\mu m^2/s$)	σ_1 (s^{-1})	σ_1' (μm)
0.28	0.6	20	0.028

σ_2 ($\mu m/s$)	σ_3 ($\mu m/s$)	σ_4 (s^{-1})	σ_4' (μm)
0.0063	0.04	0.8	0.027

form $e^{\lambda t + iqx}$. Take the parameter values used by M. HOWARD et al., 2001, listed in Table 1.1.

The equations (1.25) are valid only in a qualitative way, but establish a general basis for further improvement, which has occurred over the last years. But it should be clear that there is no general criterion to precisely decide on the form of the nonlinearities - instead, comparisons with different experiments need to be made in order to improve many details of the model.

Task. Discuss the difference between the models proposed by M. HOWARD et al., and H. MEINHARDT and P. A. J. DE BOER, both 2001.

Additional Notes

Deterministic modeling. The modelling of the dynamics of networks by nonlinear differential equations has a long history. Two of the pioneers are A. GOLDBETER and J. J. TYSON, see their work listed in the references.

The lambda phage. The lambda phage is as well a superb model system for biologists, as a system to model for the biophysicist. The first modelling attempt, based on the free energies of binding of the transcription factors and, in particular, their cooperative behaviour, is by G. ACKERS et al., 1982. Stochastic analyses were performed by A. ARKIN et al., 1998 - but this is actually a topic for the next Chapter. The octamerization of the λ repressor by looping was found by B. RÉVET et al., 1999, see also the papers by I. B. DODD et al., 2001 and 2004. Finally, the paper by J. M. G. VILAR and S. LEIBLER, 2003, extends the work by Ackers and Arkin to account for the energetics and fluctuations of looping. Papers touching on further aspects of the λ switch are by E. AURELL and K. SNEPPEN, 2002; A. BAKK et al., 2004, and M. SANTILLÁN and M. C. MACKEY, 2004.

Theory of pattern formation. *The* pioneer in the field is Alan Turing; it is worth reading his original papers and even looking into his unpublished notes which can be found in the library of the King's college in Cambridge. They are also available on the WWW. A modern classic in the theory of pattern formation is the work by A. GIERER and H. MEINHARDT, 1972.

An ongoing topic of interest (and controversy) is the robustness of the morphogen gradient in drosophila embryos. See the paper by B. HOUCH-MANDZADEH et al., 2002. A recent theory which explains the experiment is by M. HOWARD and P. R. TEN WOLDE (2005); they postulate the existence of a yet unknown corepressor.

The Min system. The Min system enjoys ongoing interest in modelling; see the papers by M. HOWARD, K. KRUSE, H. MEINHART, N. S. WINGREEN and collaborators. The system is nice since a lot can be done experimentally. Note, however, that despite the modelling approaches all being fairly similar in the underlying philosophy, there are still unanswered questions, and the modelling is not unambiguous. Thus: a problem with a lot of achievements, but still not fully solved; see the discussion by M. HOWARD and K. KRUSE, 2005. One question that remains open is, e.g., why at all an oscillation is used to determine the midplane. Other bacteria have developed stable patterns, see M. HOWARD, 2004.

Modern approaches. Min is nice, but drosophila is nicer. Relating back to the problem of transcriptional control on the chromatin level it is of enormous interest and difficulty to study the structure formation in embryo development. Modern approaches employ modelling (J. REINITZ et al., 2003), microfluidics (E. M. LUCHETTA et al., 2005), and gene network engineering (M. ISALAN et al., 2005).

References

G. K. Ackers, A. D. Johnson and M. A. Shea, *Quantitative model for gene regulation by λ phage repressor*, Proc. Natl. Acad. Sci. USA **79**, 1129-1133 (1982)

A. Arkin, J. Ross and H. H. McAdams, *Stochastic Kinetic Analysis of Developmental Pathway Bifurcation in Phage λ-Infected Escherichia coli Cells*, Genetics **149**, 1633-1648 (1998)

E. Aurell and K. Sneppen, *Epigenetics as a First Exit Problem*, Phys. Rev. Lett. **88**, 048101 (2002)

A. Bakk, R. Metzler and K. Sneppen, *Sensitivity of O_R in Phage λ*, Biophys. J. **86**, 58-66 (2004)

I. B. Dodd, A. J. Perkins, D. Tsemitsidis and J. B. Egan, *Octamerization of λ CI repressor is needed for effective repression of P_{RM} and efficient switching from lysogeny*, Genes & Development **15**, 3013-3022 (2001)

I. B. Dodd, K. E. Shearwin, A. J. Perkins, T. Burr, A. Hochschild and J. B. Egan, *Cooperativity in long-range gene regulation by the λ CI repressor*, Genes & Development **18**, 344-354 (2004)

J. L. England and J. Cardy, *Morphogen gradient from a noisy source*, Phys. Rev. Lett. **94**, 078101 (2005)

C. P. Fall, E. S. Marland, J. M. Wagner and J. J. Tyson, *Computational Cell Biology*, Springer (2002)

A. Goldbeter, *Biochemical Oscillations and Cellular Rhythms*, Cambridge University Press (1996)

A. Gierer and H. Meinhardt, *A Theory of Biological Pattern Formation*, Kybernetik **12**, 30-39 (1972)

J. B. Gurdon and P. Y. Bourillot, *Morphogen gradient interpretation*, Nature **413**, 797-803 (2001)

J. Hasty, J. Pradines, M. Dolnik and J. J. Collins, *Noise-based switches and amplifiers for gene expression*, Proc. Natl. Acad. Sci. USA **97**, 2075-2080 (2000)

B. Houchmandzadeh, E. Wieschaus and S. Leibler, *Establishment of developmental precision and proportions in the early Drosophila embryo*, Nature

415, 798-802

M. Howard, A. D. Rutenberg and S. de Vet, *Dynamic Compartmentalization of Bacteria: Accurate Cell Division in E. Coli*, Phys. Rev. Lett. **87**, 278102 (2001)

M. Howard and K. Kruse, *Cellular organization by self-organization: mechanims and models for Min protein dynamics*, J. Cell Biol. **168**, 533-536 (2005)

M. Howard, *A Mechanism for Polar Protein Localization in Bacteria*, J. Mol. Biol. **335**, 655-663 (2004)

M. Howard and P. Rein ten Wolde, *Finding the Center Reliably: Robust Patterns of Developmental Gene Expression*, Phys. Rev. Lett. **95**, 208103 (2005)

K. Huang, Y. Meir and N. S. Wingreen, *Dynamic structures in Escherichia coli: Spontaneous formation of MinE rings and MinD polar zones*, Proc. Natl. Acad. Sci. USA **100**, 12724-12728 (2003)

M. Isalan, C. Lermerle and L. Serrano, *Engineering Gene Networks to Emulate Drosophila Embryonic Pattern Formation*, PLOS Biology **3**, e64 (2005)

E. M. Luchetta, J. H. Lee, L. A. Fu, N. H. Patel and R. F. Ismagilov, *Dynamics of Drosophila embryonic patterning network perturbed in space and time using microfluidics*, Nature **434**, 1134-1138 (2005)

G. Meacci and K. Kruse, *Min-oscillations in Escherichia coli induced by interactions of membrane-bound proteins*, Phys. Biol. **2**, 89-97 (2005)

H. Meinhardt and P. A. J. de Boer, *Pattern Formation in Escherichia coli: A model for the pole-to-pole oscillations of Min proteins and the localization of the division site*, Proc. Natl. Acad. Sci. USA **98**, 14202-14207 (2001)

H. Merlitz, K. Rippe, K. v. Klenin and J. Langowski, *Looping Dynamics of Linear DNA Molecules and the Effect of DNA Curvature: A Study by Brownian Dynamics Simulation*, Biophy. J. **74**, 773-779 (1998)

M. Ptashne, *A Genetic Switch. Third Edition: Phage Lambda Revisited*, Cold Spring Harbor Laboratory Press (2004)

J. Reinitz, S. Hou and D. H. Sharp, *Transcriptional Control in Drosophila*, ComPlexUs **1**, 54-64 (2003)

B. Révet, B. von Wilcken-Bergmann, H. Bessert, A. Barker and B. Müller-Hill, *Four dimers of λ repressor bound to two suitably spaced pairs of λ operators form octamers and DNA loops over large distances*, Curr. Biology **9**, 151-154 (1999)

M. Santillán and M. C. Mackey, *Why the Lysogenic State of Phage λ Is So Stable: A Mathematical Modeling Approach*, Biophys. J. **86**, 75-84 (2004)

J. J. Tyson, K. Chen and B. Novak, *Network dynamics and cell physiology*, Nat. Rev. Mol. Cell Biol. **2**, 908-916 (2001)

J. M. G. Vilar and S. Leibler, *DNA looping and physical constraints on transcription regulation*, J. Mol. Biol. **331**, 981-989 (2003)

Chapter 2

Network Dynamics II: Fluctuations

Fluctuations are natural features of biological networks given that molecules are discrete objects and, in many circumstances in biology, only present in small numbers. To be more specific, what we want to address here are some aspects of the *functional role* fluctuations can play in biological networks. We do this by way of example, since a general theory of fluctuations in biological systems does not yet exist.[1] The examples are chosen as applications of two of the theoretical approaches discussed in the first part of the book, the Langevin-approach and the chemical master equation. The last part of this Chapter is devoted to a more general discussion of fluctuations, coming back to the distinction between *intrinsic* and *extrinsic noise* sources, this time in the context of biology.

2.1 Noise in signalling

We begin the discussion in this section by considering the dynamics associated with that of a binding site for signalling molecules, say a promoter region for a transcription factor, present at a concentration c. In Chapter 1, we had considered the binding of transcription factors at the operator sites within a model based on differential equations, in which the descriptors were the molecular concentrations. Here, we want to look a bit closer now, and understand the binding and unbinding of the transcription factors as a process leading to fluctuations in the occupancy of the binding sites. How can we characterize these fluctuations?

We address this issue first again in the framework of differential equations. The occupancy of the promoter site is, within a deterministic modeling setup, given by a kinetic equation for the time-dependent occupancy of the binding sites $n(t)$ for a given total concentration c of binding molecules, (W. BIALEK

---

[1]Is such a thing to be expected?

and S. SETAYESHGAR, 2005)

$$\dot{n} = k_+ c \left(1 - n(t)\right) - k_- n(t) \,. \tag{2.1}$$

At equilibrium, the binding is determined by detailed balance and associated with the binding free energy

$$F = k_B T \ln \left(\frac{k_+ c}{k_-} \right). \tag{2.2}$$

We now assume that thermal fluctuations affect the rate constants. We linearize eq.(2.1) around the mean value $\bar{n}$ to obtain

$$\delta \dot{n} = -(k_+ c + k_-)\delta n + c(1 - \bar{n})\delta k_+ - \bar{n}\delta k_- \,, \tag{2.3}$$

and, in addition,

$$\frac{\delta k_+}{k_+} - \frac{\delta k_-}{k_-} = \frac{\delta F}{k_B T}. \tag{2.4}$$

Together, this yields the expression

$$\delta \dot{n} = -(k_+ c + k_-)\delta n + k_+ c(1 - \bar{n})\frac{\delta F}{k_B T}. \tag{2.5}$$

We can Fourier transform the equation for $\delta n(t)$ and obtain the result (*Exercise!*)

$$\frac{\delta n(\omega)}{\delta F(\omega)} = \frac{1}{k_B T} \left[\frac{k_+ c(1 - \bar{n})}{-i\omega + (k_+ c + k_-)} \right]. \tag{2.6}$$

The total variance of the fluctuations can be obtained by invoking the *dissipation-fluctuation theorem*[2]

$$\langle (\delta n)^2 \rangle = \int \frac{d\omega}{2\pi} S_n(\omega) \tag{2.7}$$

where

$$S_n(\omega) = \frac{2k_+ c(1 - \bar{n})}{\omega^2 + (k_+ c + k_-)^2}, \tag{2.8}$$

which yields

$$\langle (\delta n)^2 \rangle = \bar{n}(1 - \bar{n}) = \frac{k_+ c}{k_+ c + k_c}(1 - \bar{n}). \tag{2.9}$$

---

[2]The DFT in this context can be written as

$$\langle n(t)n(t') \rangle = \int \frac{d\omega}{2\pi} e^{-i\omega(t-t')} S_n(\omega)$$

Note that the inverse of the sum of the rates

$$\tau_c \equiv \frac{1}{k_+ c + k_-} \tag{2.10}$$

is the correlation time of the fluctuations.

We now want to generalize this result to the case where the concentration c of transcription factors is allowed to vary spatially, i.e., it obeys a diffusion equation with a sink term,

$$\partial_t c(\mathbf{x}, t) = D\Delta c(\mathbf{x}, t) - \delta(\mathbf{x} - \mathbf{x_0})\dot{n}(t). \tag{2.11}$$

One finds as a result the generalization of eq.(2.6),

$$\frac{\delta n(\omega)}{\delta F(\omega)} = \frac{k_+ c(1 - \overline{n})}{k_B T} \left[\frac{1}{-i\omega(1 + \Sigma(\omega)) + \tau_c^{-1}} \right] \tag{2.12}$$

with

$$\Sigma(\omega) = k_+(1 - \overline{n}) \int \frac{d^3k}{(2\pi)^3} \frac{1}{[-i\omega + Dk^2]}. \tag{2.13}$$

Task. Go through the steps of the above calculation. Discuss the properties of the Fourier-space integral; it is helpful to go back to the discussion of fluctuations in the Ising model in Part I, Chapter 1. Introduce a cutoff in k-space $\Lambda = \pi/a$.

The cutoff length a is a microscopic length which we identify with the size of the binding region. If we look for the low-frequency (long-time) limit of the function $\Sigma(\omega)$, we obtain

$$\Sigma(\omega \ll D/a^2) \approx \Sigma(0) = \frac{k_+(1 - \overline{n})}{2\pi Da}, \tag{2.14}$$

and we find for the spectral density of the occupancy fluctuations the result

$$S_n(\omega) \approx 2k_+ c(1 - \overline{n}) \frac{1 + \Sigma(0)}{\omega^2(1 + \Sigma(0))^2 + \tau_c^{-2}}. \tag{2.15}$$

The allowance for spatial fluctuations in the transcription factor concentration affects the correlation time, since now

$$\tau_c^{spatial} = \frac{1 - \overline{n}}{k_-} + \frac{\langle(\delta n)^2\rangle}{2\pi Dac}. \tag{2.16}$$

The diffusion contribution adds a minimum noise level given by the root-mean-square bound per time interval Δt

$$\delta n_{rms} > \frac{\bar{n}(1 - \bar{n})}{\pi Dac\Delta t} . \qquad (2.17)$$

Let's now try to see whether this bound makes sense and plug in some numbers. In a bacterial cell, typically $N_{tf} \approx 100$ transcription factors are present in a cellular volume of 1 μm^3; their diffusion coefficient is $D \sim 3\,\mu m^2/s$. With the size of a promoter site of $a \approx 3$ nm, one finds $\pi Dc \approx 3/s$. Thus the fluctuations in the site occupancy in a time interval Δt are given by

$$\frac{\delta n}{\bar{n}} > 0.55(1 - \bar{n})\left(\frac{100}{N_{tf}}\right)^{1/2}\frac{s}{\Delta t} . \qquad (2.18)$$

Experiments on *E. coli* show that about 10 % efficiency of gene expression control can be achieved at small $\bar{n}$. In order to be consistent with the above limit, the promoter site must have an occupancy time on the order of one minute.

2.2   Stochastic cascades

In the previous section we studied how the interaction of a transcription factor with its binding site is affected by noise. In this section, we turn to the problem of how the noise that is affecting one component propagates in a network of several components. This occurs in transcriptional networks, i.e., the chain of biomolecules composed of transcription factors, DNA binding sites, mRNA products and proteins. But this is equally true also in signalling or metabolic networks, where we speak about *pathways of biochemical reactions*, where at each level fluctuation effects can intervene. One might therefore simply expect that adding level upon level of noise on top of each other might always lead to a completely washed-out output signal. So the question arises what conditions must be satisfied such that a *graded input signal* can lead to an unambiguous, e.g., all or none, output signal?

We want to address this question for a simple model for a *signalling cascade* which was proposed by M. THATTAI and A. VAN OUDENAARDEN, 2002. While motivated by real biological cascades, such as phototransduction or protein kinase cascades, the model is not tuned to a direct comparison with such systems - which would require to build in much more detail - but rather to get a general idea about the noise robustness of reaction cascades. A recent application of these modelling ideas in a study of noise propagation in gene networks has been performed by J. M. PEDRAZA and A. VAN OUDENAARDEN, 2005.

The situation we have in mind is illustrated in Figure 2.1: The cascade is

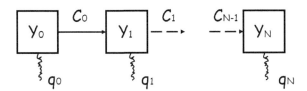

FIGURE 2.1: Generic model of a biochemical cascade, after M. THATTAI and A. VAN OUDENAARDEN, 2002. For the explanation of symbols, see text.

built up from a set of species of number Y_i, $i = 0, ..., N$ which are linked by differential amplication factors C_i which determine the response of Y_{i+1} to a change in Y_i. The input signal at Y_0 is read out at Y_N. At each level, the concentrations of the species are subject to a noise q_i. The model for this cascade that we will formulate is linear to begin with, and hence can be treated analytically using the methods of Part I, Chapter 2.

Before we write down the mathematical model for the cascade, we begin with some preliminary considerations that we will later use in the course of the discussion; they also serve as a reminder of how to deal with Langevin equations.

For the discussion, we need generic Langevin equations of the form

$$\dot{X} = g(X) + \eta(t) \tag{2.19}$$

where the noise $\eta(t)$ is defined by the moments

$$\langle \eta \rangle = 0, \quad \langle \eta(t)\eta(t + \tau) \rangle = q\delta(\tau). \tag{2.20}$$

In these equations, we understand as X the *number* of a chemical species in a cell. In the simplest case, we can assume that X is produced with a rate k_+ and destroyed with a rate k_-, where both processes are assumed Poissonian. We then have

$$\dot{X}(t) = k_+ - k_- + \eta(t) \tag{2.21}$$

and we can compute the change in X in a time interval Δt as

$$\langle \delta X \rangle = N_+ - N_- = (k_+ - k_-)\Delta t. \tag{2.22}$$

Why did we assume production and destruction of X as Poissonian? In order to make use of the relationship $\langle N_\pm^2 \rangle = \langle N_\pm \rangle$ which holds in this case. Thus we have

$$\langle \delta X^2 \rangle = (k_+ + k_-)\Delta t \,, \tag{2.23}$$

on the one hand, and, using eq.(2.20) on the other we have

$$\langle \delta X^2 \rangle = \int_0^{\Delta t} \int_0^{\Delta t} dt dt' \langle \eta(t)\eta(t') \rangle = q\Delta t \tag{2.24}$$

such that we can identify

$$q = k_+ + k_- \,. \tag{2.25}$$

We now consider a protein product Y subject to a decay rate γ. We assume that it is produced proportional to an average number b per mRNA transcript, such that the equation reads as

$$\dot{Y} = kb - \gamma Y + \eta(t) \,. \tag{2.26}$$

In a steady-state, the variance is given by

$$\langle \delta Y^2 \rangle = (1 + b)\langle Y \rangle \tag{2.27}$$

in which the factor b takes into account that the generally occurring 'bursts' in protein production will increase the variance above the Poisson level. From eq.(2.26) we find after expansion $Y \to Y_0 + \delta Y$ for δX the equation for the fluctuations

$$\delta \dot{Y} + \gamma \delta Y = \eta(t) \,. \tag{2.28}$$

By Fourier transforming we obtain

$$(i\omega + \gamma)\delta\gamma(\omega) = \eta(\omega) \,, \tag{2.29}$$

and thus in time

$$\langle |\delta Y(\omega)|^2 \rangle = \frac{q}{\omega^2 + \gamma^2} \tag{2.30}$$

and finally

$$\langle \delta Y^2 \rangle = \int_{-\infty}^{\infty} d\omega \frac{q}{\omega^2 + \gamma^2} = \frac{q}{2\gamma} \,. \tag{2.31}$$

Since $\langle Y \rangle = kb/\gamma$, we obtain $q = 2kb(1 + b)$.

After these preliminaries we can finally go on to the cascade model. Its Langevin equation reads

$$\dot{Y}_i + \gamma Y_i = F_{i-1}(Y_{i-1}) \qquad (2.32)$$

where the decay rates are all assumed equal, and we simply put them equal to one. For the fluctuations we find

$$\delta \dot{Y}_i + \delta Y_i = C_{i-1}\delta Y_{i-1} + \eta_i \qquad (2.33)$$

where the noise variance is

$$\langle \eta_i(t)\eta_i(t + \tau) \rangle = q_i \delta(\tau) . \qquad (2.34)$$

Employing the Fourier representation, we can write

$$\langle \delta Y_i^2(\omega) \rangle = \frac{q_i + C_{i-1}^2 \langle \delta Y_{i-1}^2(\omega) \rangle}{1 + \omega^2} . \qquad (2.35)$$

This equation is of the form

$$\zeta_i = \alpha_i + \beta_{i-1}\zeta_{i-1} \qquad (2.36)$$

where

$$\alpha_i \equiv \frac{q_i}{1 + \omega^2}, \quad \beta_i \equiv \frac{C_i}{1 + \omega^2}, \quad \zeta_i \equiv \langle \delta Y_i^2(\omega) \rangle . \qquad (2.37)$$

This recursion relation can be fully spelled out as

$$\zeta_N = \alpha_N + \beta_{N-1}\alpha_{N-1} + \beta_{N-1}\beta_{N-2}\alpha_{N-2} + \ldots + \beta_{N-1}\cdots\beta_0\alpha_0 . \qquad (2.38)$$

Let us look for the upper bound of the variances in the limit of an infinite cascade, i.e., for $N \to \infty$. Introducing $q \equiv \max q_i$, $C \equiv \max |C_i|$, $\alpha \equiv q/(1 + \omega^2)$ $\beta \equiv C/(1 + \omega^2)$, we have

$$\zeta_N \leq \alpha(1 + \beta + \ldots + \beta^{N-1}) + \beta\zeta_0 \qquad (2.39)$$

from which the condition

$$\zeta_\infty \leq \frac{\alpha}{1 - \beta} , \qquad (2.40)$$

and consequently

$$\langle \delta Y_\infty^2 \rangle \leq \int \frac{d\omega}{2\pi} \frac{1}{1 + \omega^2 - C^2} \leq \frac{q}{2\sqrt{1 - C^2}} \qquad (2.41)$$

follows.

The result of this calculation, eq.(2.41), shows that the output fluctuations will stay bounded provided that $|C| \leq 1$, but they can be larger than the input fluctuations at any single cascade stage, due to the presence of the factor $1/\sqrt{1 - C^2}$.

But, surprisingly, we will now see how this result can be used to attenuate the noise. For $|C| < 1$ the output noise will be independent of the input noise, if only the cascade is sufficiently long.

To demonstrate this, we again consider a cascade of species Y_i with low noise strength $q_i = q$ for all $i = 1, .., N$ which is subjected to a high noise at input, $q_0 > q$. The inverse Fourier transform of the recursion relation gives

$$\langle \delta Y_N^2 \rangle = \sum_{m=0}^{N} q_{N-m} \int \frac{d\omega}{2\pi} \frac{C^{2m}}{(1 + \omega^2)^{m+1}} = \sum_{m=0}^{N} \frac{q_{N-m}}{2} \binom{2m}{m} \left(\frac{C}{2} \right)^{2m}.$$

(2.42)

The variance of the output signal at stage N of the cascade contains contributions from the cascade stage itself, and from the input carried along. With Stirling's approximation for the factorial $m! \approx (m/e)^m \sqrt{2\pi m}$ this reads as

$$\langle \delta Y_N^2 \rangle = \frac{q}{2} \left(1 + \sum_{m=1}^{N-1} \frac{C^{2m}}{\sqrt{m\pi}} \right) + \frac{q_0}{2} \left(\frac{C^{2N}}{\sqrt{N\pi}} \right).$$

(2.43)

While the first term in this expression increases with cascade length N, the second term is attenuated exponentially, since

$$\langle \delta Y_N^2 \rangle_{input} \sim \frac{e^{-\frac{N}{N_0}}}{\sqrt{N}}$$

(2.44)

where

$$N_0 \equiv \left(- \ln(C^2) \right)^{-1}$$

(2.45)

is a sort of 'attenuation length scale'. It can serve as an estimate for the length of the cascade required to beat the noise on the input signal.

So much for a simple, linear model. The next step for us now is to see how fluctuation effects can work in systems which interact in a nonlinear fashion. We will discuss this for a - still fairly simple - feedback system in the next section.

2.3 Stochastic focusing

In this section, we discuss the stochastic dynamics of an intracellular neg-ative feedback model system (J. PAULSSON and M. EHRENBERG, 2000). In this model, two species X and S are present that regulate each other's synthe-sis. The biological motivation is taken from bacterial genomes, the so-called plasmids. Plasmids are self-replicating genomes that are able to self-control their *copy number*. We want to see how a simple system of two components can achieve this.

The macroscopic dynamic equations describing the feedback system read

$$\dot{x} = \frac{kx}{1+\alpha s} - x \tag{2.46}$$

$$\dot{s} = k_s x - k_d s \,.$$

As before, the small letters x and s denote concentration variables. The interpretation of the equations is as follows: the X-molecules multiply au-tocatalytically (like the plasmids in the motivating example), while the S molecules inhibit X-synthesis by what is called *hyperbolic inhibition*. This is reflected in the factor

$$q \equiv \frac{1}{1+\alpha s} \tag{2.47}$$

in the equation for x: the presence of a high concentration of s inhibits the production of x.

Hyperbolic inhibition is a ubiquitous control mechanism arising in vari-ous reaction schemes (see J. PAULSSON et al., 2000). In the system given by eq.(2.46), the parameters k and α set two characteristic concentration s-cales, and k_d determines how the steady state is approached: for small k_d, the approach is oscillatory, while for large values, s remains 'slaved' to x. Normal-isation of the equations with respect to the steady state values via $x_r \equiv x/\bar{x}$ and $s_r \equiv s/\bar{s}$ leads to

$$\dot{x}_r = \frac{(k-1)(1-s_r)}{1+(k-1)s_r} x_r \,, \tag{2.48}$$

$$\dot{s}_r = k_d(x_r - s_r) \,,$$

so that one immediately sees that in the limit $k \gg 1$ the rate of synthesis of X behaves has

$$\dot{x}_r \approx \left(\frac{1}{s_r} - 1 \right) x_r \,. \tag{2.49}$$

Thus, high concentrations in S favour the decrease in X, and low concentrations in S favour the increase in X. The system cannot run away to high values of X, though, since the second equation couples the evolution of X directly to that of S.

We now want to look into what will happen to the fluctuations in X and S. For this we set up the chemical master equation which corresponds to the reactions between the S and X molecules. Suppose we start with m X-molecules and n S-molecules. The synthesis of X is governed by the rate $g_{mn} = km/(1 + an)$. With the step operator $\mathcal{E}$ (see Part I, Chapter 2) one obtains the master equation in the form

$$\dot{p}_{mn} = (\mathcal{E}_m^{-1} - 1)g_{mn}\,p_{mn} + (\mathcal{E}_m - 1)m\,p_{mn} \tag{2.50}$$

$$+ k_s m(\mathcal{E}_n^{-1} - 1)\,p_{mn} + k_d(\mathcal{E}_n - 1)n\,p_{mn} + p_{mn}\sum_{n=0}^{\infty} p_{1n}$$

for $m > 0$, $n \geq 0$; the last term arises from conditioning the distribution on $m > 0$. This choice reflects the fact that the state with $X = S = 0$ is an absorbing state. X can be considered a molecule whose absence signals cell death.

We now simplify this master equation by the additional assumption that the number of S molecules rapidly adjusts to the number of X molecules; it is thus considered a fast variable and can be eliminated from the equations. The dependence on n in eq.(2.50) is thus dropped and the equation simplifies to

$$\dot{p}_m = (\mathcal{E}_m^{-1} - 1)g_m p_m + (\mathcal{E}_m - 1)m p_m + p_1 p_m. \tag{2.51}$$

This equation will now be analyzed for both noise-free and noisy signals, depending on the conditional variation of S.

When the conditional S-variation for a given value of X is negligible,

$$g_m = \frac{km}{1 + am(k_s/k_d)}. \tag{2.52}$$

If k is large (or, $am(k_s/k_d)$ small), g_m becomes constant for large m, and the number of X molecules remains Poisson distributed except at low averages $\langle m \rangle$. But when the conditional variation on S is not negligible - as a consequence of a noisy signal - eq.(2.52) has to be replaced by

$$g_m = km\sum_{n=0}^{\infty} \frac{\bar{p}_{n|m}}{1 + an}. \tag{2.53}$$

The quasistationary conditional probabilities of n S-molecules given m X-molecules $\bar{p}_{n|m}$ are Poissonian with average $\langle n \rangle_m = mk_s/k_d$, since all synthesis and degradation events are independent. The number of S-molecules at any given time represents the number of X-molecules in a probabilistic sense, and hence S is a 'noisy' slave to the slow variable X.

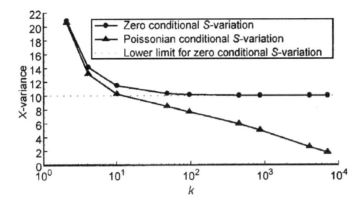

FIGURE 2.2: X-variance as a function of k for hyperbolic inhibition. Reprinted with permission from J. PAULSSON and M. EHRENBERG; Copyright (2000) American Physical Society.

Figure 2.2 shows the variance of the X-molecules as a function of k for the two cases, assuming an average of $\langle m \rangle = 10$, and $\langle n \rangle_m = m$. One sees that for the second case, the variance in X can be reduced indefinitely, while it saturates for the former. This shows that it is indeed the variations of S that lead to a reduction in the variation of X. This effect has been dubbed *stochastic focusing* - it is the paradoxical effect that the fluctuations in one component effectively reduce those in another one.

2.4 Fluctuating gene network dynamics

We have already seen in the discussion of the simple example of the λ phage how the combination of gene network elements can lead to complex dynamic behaviour. We will now go further and study a prominent example of a gene network, one which has been built artificially: the *repressilator* (M. B. ELOWITZ and S. LEIBLER, 2000). The repressilator is a gene circuit which consists of *inhibitory gates*: the binding of a transcription factor at a binding site blocks the transcription of the corresponding gene; this is similar to the binding of the λ-repressor dimer at OR3 and blocking the access for the RNA polymerase.

In this section we select a different level of description than in the previous sections: we do not want to account explicitly for the molecules binding to their binding sites, but rather prefer to treat each gene and its associated transcription and translation machinery as a composite object which we call a *gate* - in obvious analogy to the terminology common to electronic circuitry. The transcription and translation actions being combined, a gate will 'fire' when expression takes place.

In order to build a gene network, we essentially need gates with three types of actions:

- a gate which 'fires' without input, which corresponds to expression at a basal rate r in the expression of a gene;

- a gate which is blocked upon binding a transcription factor;

- a gate which fires upon input, corresponding to the binding of a transcription factor - the activated case.

All cases are illustrated schematically in Figure 2.3.

FIGURE 2.3: Three types of gene gates: a), left: constitutive transcription; b) middle: inhibition; c) right: activation of gene expression.

In order to model these gates we use a concept from theoretical computer science, a calculus called the π-*calculus* (R. MILNER, 1999). We can understand the gate actions as 'communication' events: each gate can receive input or send outputs over *communication channels*. In our case, the messages sent over the channels are simple synchronizations between output and input actions at the two different gates. The occurrence of an interaction (read: communication) requires both partners to have the corresponding input and output channels available.

In practice, this works as follows. The action of gate a) of Figure 2.3 which transcribes constitutively, will be written formally as

$$null(u) = \tau_c \,.\, (tr(u) | null(u)) \tag{2.54}$$

Note that this line should not be read as an equation but as a procedural description - as computer code. What is written on the right hand side are the computational steps that are performed.

The notation means that the gate is considered a *process* with a parameter u which is its transcription product; there is no input. The notion of process in this context differs from the one commonly used in physics: this process is an object which can perform actions in time, like a running computer program. While the notation of the π-calculus may need some time of accomodation for a statistical physicist, the use of process algebras and calculi for modelling in systems biology is increasing. A big advantage of these approaches is their modularity (or, in technical language, *compositionality*). This means that the building blocks of the networks are easy to combine and modify, much easier than is the case in the more classic approaches such as differential equations - an advantage that becomes crucial if one wants to model large systems.

Coming back to the formal description of the constitutive gate, the actions of this process are defined as follows:

- first perform a *stochastic delay* τ_c with a rate c. The rates can be determined according to the Gillespie algorithm simulating the underlying master equation of the stochastic dynamics, see Part I, Chapter 2;

- the separating dot means 'then';

- the term in brackets denotes two parallel processes x and y, $(x|y)$. The second process is a copy of the original process since processes are consumed after they have performed their actions;

- $tr(u)$ denotes the process which produces u, and has to be defined separately.

The process defining the transcription factor $tr(u)$ is written as as

$$tr(u) = !u.tr(u) + \tau_\delta.0 \tag{2.55}$$

There are two actions now among which can be chosen: $+$ denotes choice in this context. The first process is an output action (represented by the exclamation mark '!'), followed by a copy of the original process, as already discussed before. The second action is a degradation process, denoted by 0, which occurs after a delay τ_δ, where δ again is a rate.

In the same style we can now introduce the regulated gates. The gate b) of Figure 2.3 which is blocked by receiving an input (a transcription factor) can be written as

$$neg(v, u) =?v.\tau_\eta.neg(v, u) + \tau_c.(tr(u)|neg(v, u)) \qquad (2.56)$$

The second part of this expression should be clear: this is just the constitutive part. The first part means that the gate is ready to receive an input in the form of a transcription factor v which is followed by a stochastic delay; then the system returns to its initial state.

Finally the gate c) of Figure 2.3, whose expression activity is enhanced upon the action of a factor v is written as

$$neg(v, u) =?v.\tau_\eta.(tr(u)|pos(v, u)) + \tau_c.(tr(u)|neg(v, u)) \qquad (2.57)$$

The symmetry between the first and the second parts of the statement is now evident; by comparison to gate b), gate c) both produces its product u constitutively, and by activation with factor v.

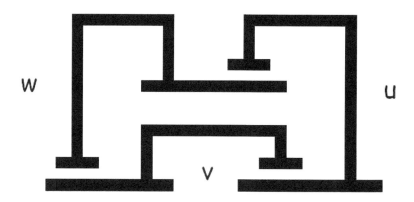

FIGURE 2.4:   The wiring diagram of the repressilator.

We now have all the elements to build the circuits. The wiring diagram of the repressilator is shown in Figure 2.4: three *neg*-gates are coupled. The wiring is expressed by the names of the products and hence the code describing the system is given by

$$neg(v, u)|neg(u, w)|neg(w, v) \qquad (2.58)$$

Figure 2.5 shows a typical simulation run, provided suitable degradation rates are chosen. As mentioned before, the underlying stochastic simulation technique is based on the Gillespie-algorithm, see Part I, Chapter 2. As was found in the experiments, the system can be made to oscillate, and all transcription factor concentrations go through clearly separated and synchronized rising and falling cycles (M. B. ELOWITZ and S. LEIBLER, 2000). The π-calculus-based simulations of the repressilator and of the more complex combinatorial gene networks built by C. C. GUET et al., 2001, have been described in R. BLOSSEY et al., 2005. An interesting result of these simulations is that stochastic effects can indeed play a role for the interpretation of the experimental data, which for some of the combinatorial circuits cannot be explained by simple logical analysis. But this is a complex issue, since another way to explain the data might be that the biological system is not completely described by the model.

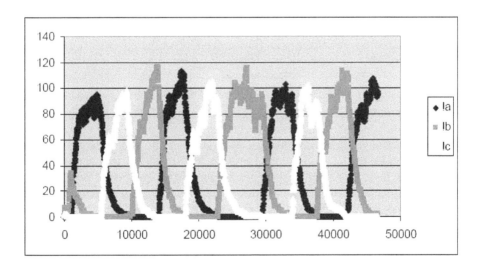

FIGURE 2.5: Repressilator oscillations. Shown are the output offers on the transcription factor channels.

2.5   Extrinsic vs. intrinsic noise

So far we have dealt with the interplay of noise in systems of few different, or many alike components. We now take a further step to distinguish more precisely the different contributions to noise, and how they can be disentangled in experiment.

P. SWAIN et al., 2002, have introduced an explicit distinction between *extrinsic* and *intrinsic* sources of noise in the context of biological systems. We have seen already before, in Part I, Chapter 2, a definition of these notions from the point of stochastic dynamic systems - a clarification is thus needed of the relation between the two concepts with the same name. Let us discuss what Swain et al. refer to with these notions.

The difference between intrinsic and extrinsic noise according to the definition by Swain et al. is easy to grasp from the following example. Consider one particular gene and its protein product - let's call it A - in a population of genetically identical cells. We want to relate this to the time evolution of the protein concentration or number in a single cell.

Even if all cells were identical at a given time, the molecular processes involved in the production of the protein (transcription and translation) are not identical in the cells; the production process of protein A will thus contain fluctuations that are intrinsic to the gene of protein A.

At the same time, the actors intervening in the production process, like the RNA-polymerase, are themselves gene products and also display cell-to-cell and time-dependent variations. Since their properties are not controlled by the production of protein A, one can consider their effect as being an extrinsic source of noise to the production of protein A. This means that the variations of the extrinsic factors (like the fluctuations in the number of polymerases) arise independently from the intrinsic variations, but clearly influence them. Obviously, extrinsic sources of noise in this sense are abundant in a cell, in which many processes are coupled.

The definition by Swain et al. is thus quite different from the one by van Kampen. In van Kampen's interpretation, systems like the ones from biology we discuss here have only intrinsic noise to begin with, since they are built up from molecular constituents - nowhere are deterministic processes in sight. The extrinsic noise by Swain et al. thus expresses the fact that the system can be separated by way of analysis way into several noise-dependent mechanisms.[3]

Let us see where this assumption leads us.

We now formulate the mathematical description for the distinction between the two types of noise suggested by P. SWAIN et al., 2002. We represent the intrinsic and extrinsic properties by two vectors $\mathbf{I}$ and $\mathbf{E}$, whose components list the different noise sources. Further, let us identify by P_k the expression level of a specific gene in cell k.

From an experimental measurement taken on an ensemble of N cells, the P_k can be deduced and averaged to obtain the moments of the corresponding protein distribution in the form

$$\frac{1}{N} \sum_{k=1}^{N} P_k^m \approx \int d\mathbf{E}\, d\mathbf{I}\, P^m(\mathbf{E}, \mathbf{I}) p(\mathbf{E}, \mathbf{I}) \tag{2.59}$$

where p is the joint probability distribution for intrinsic and extrinsic variables, and $P^m(\mathbf{E}, \mathbf{I})$ is the measured expression level for particular values of $\mathbf{E}$ and $\mathbf{I}$.

Using the product rule of probability distributions we can separate this into contributions of intrinsic and extrinsic variables by invoking conditional probabilities (see Part I, Chapter 2)

$$\frac{1}{N} \sum_{k=1}^{N} P_k^m \approx \int d\mathbf{E}\, p(\mathbf{E}) \int d\mathbf{I}\, P^m(\mathbf{E}, \mathbf{I})\, p(\mathbf{I}|\mathbf{E}), \tag{2.60}$$

and we can introduce the definition of the average over intrinsic variables from the second integral

$$\langle P^m(\mathbf{E}) \rangle \equiv \int d\mathbf{I}\, P^m(\mathbf{E}, \mathbf{I})\, p(\mathbf{I}|\mathbf{E}). \tag{2.61}$$

By contrast, averages over extrinsic variables can be indicated with an overline in the form

$$\overline{\langle P^m \rangle} = \frac{1}{N} \sum_{k=1}^{N} P_k^m. \tag{2.62}$$

---

[3]The doubling of the notions of extrinsic and intrinsic with different meanings, once in the context of general stochastic processes and once in a specific biological context, is unfortunate. In the biological context, a better notion would probably have been to talk about *cis*-noise and *trans*-noise, where cis-noise is stochasticity in the same stochastic reaction pathway, and trans-noise is that exerted from one stochastic reaction pathway onto another. Both cis- and trans-noise are intrinsic in character.

which amounts to an average over both intrinsic and extrinsic noise sources.

If we want to quantify the noise strength by standard deviations over mean, we write for the total noise, as measured in experiment

$$\eta_{tot}^2 = \frac{\frac{1}{N}\sum_k P_k^2 - \left(\frac{1}{N}\sum_k P_k\right)^2}{\left(\frac{1}{N}\sum_k P_k\right)^2}, \tag{2.63}$$

which is equivalent to

$$\eta_{tot}^2 = \frac{\overline{\langle P^2\rangle} - \left(\overline{\langle P\rangle}\right)^2}{\left(\overline{\langle P\rangle}\right)^2}. \tag{2.64}$$

This can be rewritten as

$$\eta_{tot}^2 = \frac{\overline{\langle P^2\rangle - \langle P\rangle^2}}{\left(\overline{\langle P\rangle}\right)^2} + \frac{\overline{\langle P\rangle^2} - \left(\overline{\langle P\rangle}\right)^2}{\left(\overline{\langle P\rangle}\right)^2} = \eta_{int}^2 + \eta_{ext}^2. \tag{2.65}$$

In order to distinguish experimentally between these two different contributions, one has to measure $\overline{\langle P\rangle^2}$. This quantity can be obtained as follows. If two identical copies of a particular gene were present in the same cell k, their expression levels were $P_k^{(1)}$ and $P_k^{(2)}$. These would have the same extrinsic sources of noise, but different intrinsic variables. Consequently,

$$\frac{1}{N}\sum_{k=1}^N P_k^{(1)} P_k^{(2)} \approx \int d\mathbf{E} \int d\mathbf{I_1} d\mathbf{I_2} P(\mathbf{E}, \mathbf{I_1}) P(\mathbf{E}, \mathbf{I_2}) p(\mathbf{I_1}, \mathbf{I_2}, \mathbf{E})$$

$$= \int d\mathbf{E}\, p(\mathbf{E}) \left[\int d\mathbf{I}\, P(\mathbf{E}, \mathbf{I}) p(\mathbf{I}|\mathbf{E})\right]^2$$

$$= \overline{\langle P\rangle^2}. \tag{2.66}$$

The verification of this scenario can be made by fluorescent tagging of two genes controlled by identical regulatory sequences. These experiments have been performed by M. B. ELOWITZ et al., 2002, in *E. coli*. In the absence of intrinsic noise, the concentrations of both proteins, as measured by their fluorescence, would fluctuate in a correlated way. In a population of cells this then gives rise to a certain distribution of protein levels due to extrinsic noise alone. In the presence of intrinsic noise, the correlations between the two protein concentrations in each single cell decrease, reflecting different expression levels for the proteins.

Figure 2.6 shows the experimental results in the form of a plot of noise levels as a function of a measure for population-averaged transcription rates, distinguishing between extrinsic and intrinsic noise contributions.

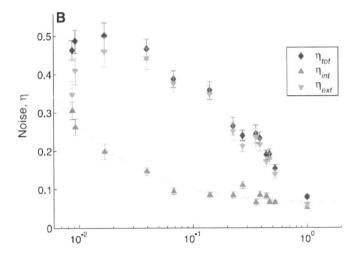

FIGURE 2.6: Cellular genetic noise in *E. coli*. Total, intrinsic and extrinsic noise as a function of fluorescence levels (population means) characterizing transcription rates. Reprinted with permission from Science from M. B. ELOWITZ et al., 2002.

Additional Notes

Fluctuations, noise and robustness (see next Section) are becoming topics of increasing interest in the computational biology community. This is to a large part, but not exclusively, due to the influx of statistical physicists into the field. An early contribution is by H. H. MCADAMS and A. ARKIN, 1997.

The effects of noise in the standard model of gene expression (DNA → RNA → protein) has been modelled and discussed with respect to experimental work in particular by J. PAULSSON (2004, 2005). Similar work is due to M. THATTAI and A. VAN OUDENAARDEN, 2001, and by A. BECSKEI, B. B. KAUFMANN and A. VAN OUDENAARDEN, 2005.

Pi-calculus and process algebras. There is a whole literature developing on systems biology modelling based on approaches by theoretical computer scientists, in particular by specialists on programming languages. The seminal work is by A. REGEV, 2002.

Intrinsic vs. extrinsic noise. This topic enjoys growing interest in the bio-community. A recent application of this concept is by A. COLMAN-LERNER et al., 2005, to the cell-fate decision system.

Parameter problems. A very general modelling problem in systems biology is the lack of knowledge of parameters - a problem of which already A. Turing was well aware. A paper on statistical mechanics approaches to systems with many poorly known parameters is by K. S. BROWN and J. P. SETHNA, 2003.

References

A. Becskei, B. B. Kaufmann and A. van Oudenaarden, *Contributions of low molecule number and chromosomal positioning to stochastic gene expression*, Nat. Genet. **37**, 937-944 (2005)

W. Bialek and S. Setayeshgar, *Physical limits to biochemical signalling*, Proc. Natl. Acad. Sci. USA **102**, 10040-10045 (2005)

R. Blossey, L. Cardelli and A. Phillips, *A compositional approach to the stochastic dynamics of gene networks*, Transactions in Computational Systems Biology IV, 99-122 (2006)

K. S. Brown and J. P. Sethna, *Statistical mechanical approaches to models with many poorly known parameters*, Phys. Rev. E **68**, 021904 (2003)

A. Colman-Lerner A. Gordon, E. Serra, T. Chin, O. Resnekov, D. Endy, C. G. Pesce and R. Brent, *Regulated cell-to-cell variation in a cell-fate decision system*, Nature **437**, 699-706 (2005)

M. B. Elowitz and S. Leibler, *Synthetic gene oscillatory network of transcriptional regulators*, Nature **403**, 335-338 (2000)

M. B. Elowitz, A. J. Levine, E. D. Siggia and P. S. Swain, *Stochastic Gene Expression in a Single Cell*, Science **297**, 1183-1186 (2002)

C. C. Guet, M. B. Elowitz, W. Hsing, S. Leibler, *Combinatorial Synthesis of Genetic Networks*, Science **296**, 1466-1470 (2002)

H. H. McAdams and A. Arkin, *Stochastic mechanisms in gene expression*, Proc. Natl. Acad. Sci. **94**, 814-819 (1997)

R. Milner, *The π-calculus*, Cambridge University Press (1999)

J. Paulsson, O. Berg and M. Ehrenberg, *Stochastic focusing: Fluctuation-enhanced sensitivity of intracellular regulation*, Proc. Natl. Acad. Sci. USA **97**, 7148-7153 (2000)

J. Paulsson and M. Ehrenberg, *Random Signal Fluctuations Can Reduce Random Fluctuations in Regulated Components of Chemical Regulatory Networks*, Phys. Rev. Lett. **84**, 5447-5450 (2000)

J. Paulsson, *Summing up the noise in gene networks*, Nature **427**, 415-418 (2004)

J. Paulsson, *Models of stochastic gene expression*, Physics of Life Reviews **2**, 157-175 (2005)

J. M. Pedraza and A. van Oudenaarden, *Noise propagation in Gene Networks*, Nature **307**, 1965-1969 (2005)

A. Regev, *Computational Systems Biology: A Calculus for Biomolecular Knowledge*, PhD-Thesis University of Tel Aviv (2002)

P. S. Swain, M. B. Elowitz and E. D. Siggia, *Intrinsic and extrinsic contributions to stochasticity in gene expression*, Proc. Natl. Acad. Sci. USA **99**, 12795-12800 (2002)

M. Thattai and A. van Oudenaarden, *Intrinsic noise in gene regulatory networks*, Proc. Natl. Acad. Sci. USA **98**, 8614-8619 (2001)

M. Thattai and A. van Oudenaarden, *Attenuation of Noise in Ultrasensitive Signaling Cascades*, Biophys. J. **82**, 2943-2950 (2002)

Chapter 3

Networks: Structure

3.1 Networks as graphs

After our study of biological networks on a small scale - i.e., network components - we now ask how a whole *biological network* can be defined and characterized. The starting point of our discussion of network structure is the random network, or *random graph* introduced by ERDÖS AND RENYI, 1959, which will serve us as a convenient null model.

The ER-graph consists of *n* *nodes* or *vertices*, joined by *links* or *edges* between the vertices which are placed at random with independent probability *p*; an illustration is given in Figure 3.1. The construction allows to introduce the notion of an *ensemble of graphs* of *n* vertices, $\mathcal{G}(n, p)$, in which each graph is present with the probability corresponding to the number of its edges.

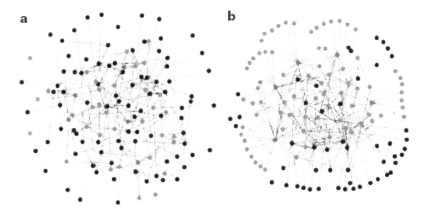

FIGURE 3.1:   Random network, or Erdös-Renyi graph a). The graph under b) is a represents a scale-free network, to be discussed in the text below. Reprinted with permission from Macmillan Publishers Ltd., R. ALBERT et al., 2000.)

The random graph has the following basic properties:

- The *degree* of a vertex is the number of edges connected to it. The *average degree* of a graph is then given by

$$z = 2 \times \frac{n(n-1)p}{2n} = (n-1)p \tag{3.1}$$

where the factor of two takes into account that each edge is connected to *two* vertices. Hence there are no edges going out and ending at the same vertex.

- The probability p_k that a vertex in an Erdös-Renyi graph has degree k is given by the binomial distribution

$$p_k = \binom{n-1}{k} p^k (1-p)^{n-1-k} \tag{3.2}$$

which in the limit $n \gg kz$ yields a Poisson distribution

$$p_k = \frac{z^k}{k!} e^{-z} . \tag{3.3}$$

The degree distribution is thus strongly peaked around the mean z, and decays rapidly for large degrees.

Exercise. Show that ER-graphs are always *treelike*. A graph is called treelike if each of its cycles (or circuits: say internal loops) has at most one common vertex.

The question can be posed whether the random graph model applies to biological systems (or to other existing network-like systems, for that matter). An attempt in this direction was made by H. JEONG et al., 2000, in the context of a study of metabolic networks of 43 different organisms. A *metabolic network* is the network of biochemical compounds whose transformations are needed to sustain body function. It was generally found that these large-scale networks display asymmetric degree distributions with rather 'fat' tails, and could be characterized by algebraic behaviour for a range of degrees, k. In analogy to critical phenomena - think of the algebraic behaviour of correlation functions near a critical point discussed in Part I, Chapter 1, such networks have been dubbed '*scale-free*'.

In subsequent work, H. JEONG et al., 2001, considered the interaction network built from proteins in yeast *Saccharomyces cerevisiae*, consisting of 1870 edges and 2240 identified physical interactions among the proteins. Their finding is shown in Color Figure 8. Again, the degree distribution displays a fat tail, and the authors fit it to the expression

$$p(k) = \frac{A}{(k + k_D)^\gamma} \exp(-k/k_c) \tag{3.4}$$

where $k_D \approx 1$ and $k_c \approx 20$ are short- and large scale cutoffs.

Evidently, the presence of cutoffs shows that the network is *not* strictly scale-free. This notion has indeed been much abused in the literature, and we will come back to it below in more detail. At present, it is important to also add that the algebraic behaviour is observed only over a quite limited range of k-values, sometimes hardly more than one order of magnitude. The exponent γ is quite typically found to line in the range $2 < \gamma < 3$.

Independently of the issue whether the graph is truly scale-free or not, already the appearance of a fat tail in the degree distribution obviously takes the protein interaction network out of the basic random graph paradigm. As a consequence, in order to understand these networks we need new theory. Some additional observations provide further indication what this new theory should be able to deliver.

The protein network is an example of a network which is undirectional, i.e. the interaction of proteins does not distinguish between the partners. This does not hold for all networks: e.g., we may want to study a network of transcriptional interactions. In such a network, interactions are directional. A transcription factor will activate a specific gene (we have seen in Chapter 1 how this works), such that we need to have edges represent not just a basic 'interacts with', but also edges which are directed from one node to another. A typical subgraph is indicated schematically in Figure 3.2.

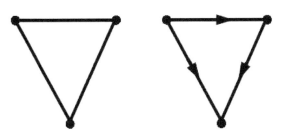

FIGURE 3.2: An undirected and a directed graph.

The fact that the construction of networks as graphs must reflect the particular identity of the interacting components explains immediately why the knowledge of degree distributions is not enough to characterize them. For directed networks, we have to distinguish between ingoing and outgoing degrees; but there are probably many deeper levels of structure as well. Also, it may not be sufficient to look at only one type of network (it may be selected only because of the availability of data): S. MASLOV and K. SNEPPEN, 2002, have shown that the connectivity of protein interaction networks and transcriptional regulation networks in the yeast *Saccharomyces cerevisiae* show important correlations. The size of the networks used in this study amount to 4549 physical interactions (edges) between 3278 proteins (vertices) for the protein network, and 1289 positive or negative direct transcriptional interactions for 682 proteins.

In order to go beyond the ER-graph-network paradigm, we first list a number of important network characteristics and then show how to compute them from a given degree distribution by the method of generating functions. We then move on to a more general statistical mechanical theory of networks.

3.2   Probability generating functions and network characteristics

Probability generating functions. We have encountered generating functions in previous Chapters, but only in this application they will rise to the height of their powers. We begin by taking the vertex degree distribution of the random network introduced before, p_k, and define, following (M. E. J. NEWMAN, 2003), the generating function

$$G_0(x) = \sum_{k=0}^{\infty} p_k x^k . \tag{3.5}$$

The vertex degree distribution p_k can be generated from G_0 by differentiation,

$$p_k = \frac{1}{k!} \frac{d^k G_0}{dx^k}\bigg|_{x=0} . \tag{3.6}$$

G_0 has the following properties:

1. If G_0 is properly normalized, then

$$G_0(1) = \sum_k p_k = 1. \tag{3.7}$$

2. The average degree $\langle k \rangle$ is found from computing

$$\frac{dG_0(x)}{dx}\bigg|_{x=1} \equiv G_0'(1) = \sum_k k p_k = \langle k \rangle. \tag{3.8}$$

3. For the n-th moment of p_k one has

$$\langle k^n \rangle = \sum k^n p_k = \left[\left(x \frac{d}{dx} \right)^n G_0(x) \right]_{x=1}. \tag{3.9}$$

We may of course also consider generating functions of probability distributions other than the degree distribution. E.g., the probability that a vertex that can be reached along one edge from a starting vertex has $(k-1)$ outgoing edges is given by (*Exercise!*)

$$q_{k-1} = k \frac{p_k}{\sum_j j p_j}. \tag{3.10}$$

The average degree of the vertex then follows as

$$\sum_{k=0}^{\infty} k q_k = \frac{\langle k^2 \rangle - \langle k \rangle}{\langle k \rangle}. \tag{3.11}$$

This is the average degree of a vertex two edge distances away from the starting vertex. Based on this result, we can define the number of second neighbors in the graph by

$$z_2 = \left(\sum_k k q_k \right) \cdot \langle k \rangle = \langle k^2 \rangle - \langle k \rangle. \tag{3.12}$$

The probability generating function based on q_k finally reads as

$$G_1(x) = \sum_{k=0}^{\infty} q_k x^k = \frac{G_0'(x)}{z_1} \tag{3.13}$$

where $z_1 \equiv \langle k \rangle$.

We can now list further important network measures:

- The *clustering coefficient* C of a random graph is given by

$$C \equiv \frac{1}{nz} \left[\sum_k k q_k \right]^2 = \frac{z}{n} \left[c_v^2 + \frac{z-1}{z} \right]^2 \tag{3.14}$$

  where c_v is the *coefficient of variation* of the degree distribution (the ratio of standard deviation to mean).

- The *diameter d of a graph* is given by the maximum distance between any two connected vertices in the graph. It can be shown that the fraction of all possible graphs with n vertices and m edges for which $d \geq c \ln n$ for some constant c tends to zero in the limit $n \to \infty$.

- A *giant component* is a subgraph whose size scales linearly with the graph size. It corresponds to the *infinite cluster* in percolation theory.[1] The random graph has a phase transition (akin to the percolation transition on a lattice) in which a giant component is present whenever the *Molloy-Reed criterion* holds

$$\sum_{k=0}^{\infty} k(k-2)p_k = 0 \,. \tag{3.15}$$

- For networks which contain a giant component the average vertex-vertex distance ℓ is given by

$$\ell = \frac{\ln(n/z_1)}{\ln(z_2/z_1)} + 1 \,. \tag{3.16}$$

Task. Characterize correlations in the ER-graph.

Robustness. The degree distribution allows to characterize networks according to their *robustness* or resilience to attack. R. ALBERT et al., 2000, compared the properties of an ER-graph and a scale-free graph upon edge removal. These networks differ in a characteristic way: since vertices and edges in ER-graphs are more 'alike', its properties such as the average degree change fairly little when edges are removed randomly or even on purpose, i.e., when vertices with a high degree are removed preferentially, simulating an attack on the network. By contrast, the removal of few, highly connected nodes in a scale-free network affects its structure very much.

Exercise. Study the robustness of the network described by the generating function

$$G_0(x) = \frac{1}{n}\left((n-1)x^3 + x^n\right) \,. \tag{3.17}$$

Picture this network. How does its diameter change as a function of n?

---

[1] Percolation is a geometric phase transition: consider the case of a two-dimensional regular lattice with all sites occupied. Now one places bonds between the sites with a given probability. A question one can pose now is at what concentration of occupied sites will a cluster arise which spans the whole lattice: this is the infinite cluster. Its appearance (the percolation transition) and the statistical properties of this cluster share many features of thermal phase transitions, but there is of course no free energy in this case.

We are now ready for a more systematic approach, along the lines of e-quilibrium statistical mechanics of the first Chapter of Part I. As there, the notion of entropy proves to be highly useful.

3.3 Statistical mechanics of networks

A statistical mechanical theory of networks can be developed for *exponential random graphs* (J. PARK and M. E. J. NEWMAN, 2004). If G is an element of a set of graphs $\mathcal{G}$, we can obtain its probability distribution $P(G)$ by maximizing the Gibbs entropy

$$S = -\sum_{G \in \mathcal{G}} P(G) \ln P(G) \tag{3.18}$$

under the constraints $\sum_G P(G) x_i(G) = \langle x_i \rangle$ and $\sum_G P(G) = 1$, where x_i, $i = 1, ..., r$ are a set of graph observables. The derivation of $P(G)$ can now be performed in complete analogy to our discussion of the thermal ensembles in Part I (*Task!*), leading to the result

$$P(G) = \frac{e^{-\sum_i \theta_i x_i(G)}}{Z} = \frac{e^{-H(G)}}{Z}. \tag{3.19}$$

As in the thermal case, the expression eq.(3.19) involves a set of Lagrange multipliers θ_i.

Examples. The simplest example is the random graph with a fixed number of vertices n. If we want to characterize the graph only by the mean number of edges, $\langle m \rangle$, the Hamiltonian in eq.(3.19) is chosen to be

$$H(G) = \theta m(G) \tag{3.20}$$

We can now evaluate the partition function Z for an ensemble of simple undirected graphs on n vertices. An $(n \times n)$ adjacency matrix can be defined by

$$\sigma_{ij} = \begin{cases} 1 & i \leftrightarrow j, \\ 0 & else \end{cases} \tag{3.21}$$

where the double-arrow symbol $\leftrightarrow$ denotes i connected to j. With the adjacency matrix we can represent the number of edges as

$$m = \sum_{i<j} \sigma_{ij}, \tag{3.22}$$

and hence compute the partition function Z

$$Z = \sum_G e^{-H} = \sum_{\{\sigma_{ij}\}} \exp\left(-\theta \sum_{i<j} \sigma_{ij}\right)$$

$$= \prod_{i<j} \sum_{\sigma_{ij}=0}^{1} e^{\theta \sigma_{ij}} = \prod_{i<j}(1 + e^{-\theta}) = [1 + e^{-\theta}]^{\binom{n}{2}} \tag{3.23}$$

We can also define the analogue of the free energy by $F \equiv -\ln Z$ and find the result

$$F = -\binom{n}{2}\ln(1 + e^{-\theta}), \tag{3.24}$$

such that the expected number of edges in the graph is

$$<m> = \frac{1}{Z}\sum_G me^{-H} = \frac{\partial F}{\partial \theta} = \binom{n}{2}\frac{1}{1 + e^{\theta}}. \tag{3.25}$$

In the last equation, we may want to redefine $p = (1 + e^{\theta})^{-1}$ to simplify notation. The probability of a graph in this ensemble is then given by

$$P(G) = \frac{e^{-H}}{Z} = p^m(1 - p)^{\binom{n}{2} - m}. \tag{3.26}$$

$P(G)$ is the probability for a graph in which each of the $\binom{n}{2}$ possible edges appears with probability p; this is just a Bernoulli random graph.

Further examples. We list some further examples; the computation of their properties is left as an *Exercise*. (J. PARK and M. E.J. NEWMAN, 2004).

Specifying degrees. If we choose the vertex degrees k_i as observables,

$$H = \sum_i \theta_i k_i, \tag{3.27}$$

we can rewrite this using the adjacency matrix σ_{ij} as

$$H = \sum_{ij} \theta_i \sigma_{ij} = \sum_{i<j}(\theta_i + \theta_j)\sigma_{ij}. \tag{3.28}$$

The partition function is given by

$$Z = \prod_{i<j}(1 + e^{-(\theta_i + \theta_j)}). \tag{3.29}$$

Directed graphs. We now change the graph ensemble $\mathcal{G}$ to contain directed graphs; we have to adopt the adjacency matrix accordingly. It now contains an entry 1 if an edge exists in the graph from j to i ($j \to i$). For the choice $H = \theta m$ one has

$$Z = \prod_{i \neq j} \sum_{\sigma_{ij}=0}^{1} e^{-\theta \sigma_{ij}} = [1 + e^{-\theta}]^{2\binom{n}{2}} \tag{3.30}$$

Fixed edges. If one chooses for $\mathcal{G}$ the set of graphs with both fixed number of vertices n and edges m,

$$Z = \sum_{G} \delta(\tilde{m}, m(G)) e^{-H} \tag{3.31}$$

where $\tilde{m}$ is the desired number of edges. This construction corresponds to a canonical ensemble (J. BERG and M. LÄSSIG, 2002).

3.4 Network growth

We have now learnt how to characterize statistically graphs or networks of more general nature than simple random graphs, so that we feel ready to come back to the problem of protein network structure. We can determine, e.g., degree distributions empirically based on data and characterize the networks by various measures. A recent detailed analysis by J.-D. H. HAN et al., 2005, demonstrated the pitfalls one can get into by an uncritical data analysis: the limits on the sampling range alone can already lead to wrong conclusions on the degree distributions.

Whatever we do with the existing incomplete data, this does not, clearly, tell us how they came about in the first place; it is hence of interest to model the evolution of a network. For the protein-protein interactions networks, e.g., known mechanisms of network evolution are *gene duplication* and *mutation*. In order to build a simple model for the dynamics of network growth then two basic assumptions can be made (R. ALBERT and A.-L. BARABÁSI, 2000; J. KIM et al., 2002):

- **Vertex duplication.** Vertices (i.e., new proteins) are added, one after the other. A new vertex duplicates a previously existing vertex which is chosen randomly, and links to its neighbors are placed with a probability $1 - \delta$.

- **Diversification.** Each new vertex links to any previous node with probability β/n, where n is the current total number of vertices in the network.

A protein interaction network generated from such mechanisms thus mimics the underlying evolutionary mechanisms. In these, mutations occur both at the duplication and diversification levels, if the parameters $\beta, \delta > 0$.

The average vertex degree of such a network G can be estimated as follows. In each growth step, the average number of edges $\langle m \rangle$ increases by $\beta + (1-\delta)G$. Thus,

$$\langle m \rangle = (\beta + (1 - \delta)G)n \, . \tag{3.32}$$

Since generally the relation $G = 2\langle m \rangle/n$ holds, we find, eliminating $\langle m \rangle$

$$G = \frac{2\beta}{2\delta - 1} \, , \tag{3.33}$$

which can only hold for $\delta > \delta_c = 1/2$. Below this threshold, the number of links grows according to

$$\frac{d\langle m \rangle}{dn} = \beta + 2(1 - \delta)\frac{\langle m \rangle}{n} \tag{3.34}$$

which together with the relation $G(n) = 2\langle m \rangle(n)/n$ yields the scaling dependencies

$$G(n) = \begin{cases} finite & \delta > 1/2 \, , \\ \beta \ln n & \delta = 1/2 \, , \\ C \times n^{1-2\delta} & \delta < 1/2 \, . \end{cases} \tag{3.35}$$

Without diversification ($\beta = 0$), a finite average vertex degree is therefore only found if $\delta > 1/2$, illustrating the important role mutations play.

The case of $\delta > 1/2$, $\beta > 0$ can be studied in more detail by the rate equation for the evolution of the number of vertices of degree k, $P_k(n)$ when the network as a whole has n vertices. The degree of a node increases by one at a rate $a_k = \beta + (1-\delta)k$. We can then write down a rate equation of the form

$$\frac{dP_k(n)}{dn} = \frac{1}{n}(a_{k-1}P_{k-1} - a_k P_k) + G_k \tag{3.36}$$

where the first two terms account for the increase of a vertex degree by one. The last term is a source term for the introduction of new vertices with k edges, with a of the edges created by duplication and $b = k - a$ by diversification. The probability of the duplication process is

$$g_{dup} = \sum_{s \geq a} x_s \binom{s}{a} (1 - \delta)^a \delta^{s-a} \tag{3.37}$$

where $p_s = n_s/n$ is the probability of a vertex of degree s chosen for the duplication process. The probability of diversification is

$$g_{div} = \beta^b \frac{e^{-\beta}}{b!} \tag{3.38}$$

such that the full expression for G_k is given by

$$G_k = \sum_{a+b=k} \sum_{s=a} x_s \binom{s}{a} (1 - \delta)^a \delta^{s-a} \beta^b \frac{e^{-\beta}}{b!}. \tag{3.39}$$

Since the n_k grow linearly in n, we can plug the relation $P_k(n) = np_k$ into the rate equation and obtain

$$\left(k + \frac{1 + \beta}{1 - \delta} \right) p_k = \left(k - 1 + \frac{\beta}{1 - \delta} \right) p_{k-1} + \frac{G_k}{1 - \delta}, \tag{3.40}$$

which looks very much like a recursion relation for the p_k - but it isn't, since G_k depends on all p_s for $s \geq k$. It can, however, be turned into a recursion relation in the limit $k \to \infty$. In this limit, the main contribution to G_k arises for small values of b, and the summand is sharply peaked around $s \approx k/(1 - \delta)$. We may then replace the lower limit by $s = k$ and p_s by its value at $s = k/(1 - \delta)$. Further, anticipating that p_k decays as $k^{-\gamma}$ we introduce the *ansatz* $p_s = (1 - \delta)^\gamma p_k$ and finally simplify G_k according to

$$G_k \approx (1 - \delta)^\gamma p_k \sum_{s=k} x_s \binom{s}{k} (1 - \delta)^k \delta^{s-k} \sum_{b=0}^{\infty} \beta^b \frac{e^{-\beta}}{b!} = (1 - \delta)^{\gamma-1} p_k \tag{3.41}$$

since the binomial sum equals $(1 - \delta)^{-1}$.

Summing this all up, in the limit $k \to \infty$, p_k is found to have a power-law behaviour $p_k \sim k^{-\gamma}$ with the value of γ fixed by the relation

$$\gamma(\delta) = 1 + \frac{1}{1 - \delta} - \frac{1}{(1 - \delta)^{2-\gamma}} \tag{3.42}$$

We thus find that $\gamma(\delta)$ sensitively depends on δ, but *not* on β. Choosing a value of $\delta = 0.53$, as can be suggested from observations (A. WAGNER, 2001), a value of $\gamma = 2.373..$ is found, in accord with what we had indicated before.

A particularly important point to notice is that this result is an asymptotic one. The convergence to the asymptotic limit in which the power laws reign is very slow. Even for vertex numbers of $n = 10^6$, the power-law regime is only reached over two orders of magnitude of degrees k. Compare this with the data for real protein-networks...

Additional Notes

The literature on networks emerging within the statistical physics community has literally exploded. An early review was written by R. ALBERT AND A. L. BARABÁSI, 2002. The mathematical theory of random graphs in the tradition of the pioneers P. ERDÖS and A. RENYI, 1959, is summarized by B. BOLLOBÁS, 2000.

Small worlds. Another important network paradigm that has been uncovered in the last years is that of *small-world networks* (D. J. WATTS and S. H. STROGATZ, 1998). The authors studied a random rewiring process of a completely regular network. Initially, the model network is placed on a one-dimensional ring and up to k neighbouring points are linked. Then links are broken and rewired at random across the lattice. Upon the introduction of only a few of these 'shortcuts', a significant clustering property of the network arises, characterizing an intermediate stage between complete regularity on the one side, and complete randomness on the other.

The important feature of the small-world model is that distant points on the network get access to each other: only a small number number of short-cuts will thus allow the network, although being still fairly sparse, to 'communicate' efficiently across all nodes. This phenomenon has acquired some fame through the famous story of 'six degrees of separation', referring to the number of acquaintances needed to pass between two randomly selected people in the world. M. E. J. NEWMAN at al., 2000, formulated an exactly solvable mean-field model for the small-world network, but despite being 'just a mean-field solution', it is already quite demanding.

Motifs and modules. The educated reader might wonder why the notion of 'motifs' was not discussed in the main text. Well, we make up for that now a little bit. For details, the reader is asked to consult the book by U. ALON, 2006.

Biological networks may display particular global features, but they will also be characterized by the properties of local neighbourhoods of vertices. This has been noticed via the occurrence of particular *network motifs* (R. MILO et al., 2002, S. S. SHEN-ORR et al., 2002).

Figure 3.3 shows a prominent example, the *feedforward loop*.[2] While in a random graph subgraphs are more likely to be trees, functional subgraphs of networks are more likely to have a higher number of edges than a subgraph in a random network. Again the distinction between directed and undirected

---

[2]Note that the innocently looking graph to the right in Fig. 3.2 has the same topology!

graphs matters.

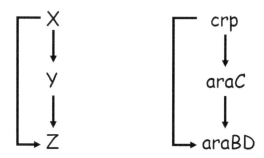

FIGURE 3.3:   A feedforward loop. Left: basic structure of the loop; right: an example of a feedback loop in the gene network of *E. coli*, after S. S. SHEN-ORR, 2002.

There have been meanwhile a number of studies focussing on network motifs, e.g. for *E. coli* and *S. cerevisiae* (E. YEGER-LOTEM et al., 2004). In particular the feedforward loop has interesting features which have been recently discussed (S. MANGAN and U. ALON, 2003; S. KALIR et al., 2005). Eight different types of feedforward loops have been identified, depending on the actions of the transcription factors (activating or repressing).

In *E. coli*, aside from the feedforward loop, two other motifs have been detected as being statistically overrepresented in the data: the so-called single input module, in which a single transcription factor regulates a set of transcriptional units, the operons, see our discussion of the phage lambda. A third motif is the *dense overlapping regulon*, in which a number of operons is regulated by a set of transcription factors.

A review of the different types of motifs that have been detected is by D. M. WOLF and A. A. ARKIN, 2003). The notion of a motif generalizes, on a higher network level, to that of a *network module*, comprising as separately functioning network elements (E. RAVASZ et al., 2002). The distinction between a motif and a module is not evident, as Wolf and Arkin note themselves.

The existence of motifs can be used to construct an algorithmic procedure for the detection and significance decision of local network elements. The deterministic motifs are generalized to probabilistic motifs in which edges occur with a certain likelihood. This procedure, called *graph alignment* due to its analogy to sequence alignment, can be based on a suitably defined *scoring function*, i.e., a Hamiltonian for the motifs, H_{score} (BERG AND LÄSSIG, 2004).

Finally, it should be stressed that the overrepresentation of certain motifs does not at all guarantee that these are the important building blocks in the system. Biological networks, after all, have evolved over the course of many thousands of years, and the networks hence carry a history along which is very difficult to take into account in the present description of networks (see, e.g., E. DEKEL et al., 2005, on the selection of the feedforward motif). As useful as such notions as motifs and modules are as a work guide at present (for a discussion, see the paper by L. H. HARTWELL et al., 1999), as little is known about their true biological significance.

References

R. Albert, H. Jeong and A.-L. Barabási, *Error and attack tolerance of complex networks*, Nature **406**, 378-380 (2000)

R. Albert and A.-L. Barabási, *Statistical mechanics of complex networks*, Rev. Mod. Phys. **74**, 47-97 (2002)

U. Alon, *Introduction to Systems Biology and the Design Principles of Biological Networks*, Chapman & Hall/CRC (2006)

A.-L. Barabási and R. Albert, *Emergence of Scaling in Random Networks*, Science **286**, 509-512 (1999)

J. Berg and M. Lässig, *Correlated Random Networks*, Phys. Rev. Lett. **89**, 228701 (2002)

J. Berg and M. Lässig, *Local graph alignment and motif search in biological networks*, Proc. Natl. Acad. Sci. USA **101**, 14689-14694 (2004)

B. Bollobás, *Random Graphs*, Academic (2000)

E. Dekel, S. Mangan and U. Alon, *Environmental selection of the feed-forward loop circuit in gene-regulation networks*, Phys. Biol. **2**, 81-88 (2005)

P. Erdös and A. Renyi, On random graphs, Publ. Math. Debrecen, **6**, 290-297 (1959)

J.-D. J. Han, D. Dupuy, N. Bertin, M. E. Cusick and M. Vidal, *Effect of sampling on topology predictions of protein-protein interaction networks*, Nat. Biotech. **23**, 839-844 (2005)

J. H. Hartwell, J. J. Hopfield, S. Leibler and A. W. Murray, *From molecular to modular cell biology*, Nature **402**, C47-C50 (1999)

H. Jeong, B. Tombor, R. Albert, Z. N. Oltvai and A.-L. Barabási, *The large-scale structure of metabolic networks*, Nature **407**, 651-654 (2000)

H. Jeong, S. P. Mason, A.-L. Barabási and Z. N. Oltvai, *Lethality and centrality in protein networks*, Nature **411**, 41-42 (2001)

S. Kalir, S. Mangan and U. Alon, *A coherent feed-forward loop with a SUM input function prolongs flagella expression in Escherichia coli*, Mol. Syst. Biol., msb44100010 (2005)

J. Kim, P. L. Krapivsky, B. Kahng and S. Redner, *Infinite-Order Percolation and Giant Fluctuations in a Protein Interaction Network*, Phys. Rev. E **66**, 055101 (2002)

S. Mangan and U. Alon, *Structure and function of the feed-forward loop network motif*, Proc. Natl. Acad. Sci. USA **100**, 11980-11985 (2003)

S. Maslov and K. Sneppen, *Specificity and Stability in Topology of Protein Networks*, Science **296**, 910-913 (2002)

R. Milo, S. S. Shen-Orr, S. Itzkovitz, N. Kashtan, D. Chklovskii and U. Alon, *Network Motifs: Simple Building Blocks of Complex Networks*, Science **298**, 824-827 (2002)

M. E. J. Newman, *Random graphs as models for networks*, in *Handbook of Graphs and Networks*, S. Bornholdt and H. G. Schuster (eds.), (2003)

M. E. J. Newman, C. Moore and D. J. Watts, *Mean-field Solution of the Small-World Network Model*, Phys. Rev. Lett. **84**, 3201-3204 (2000)

J. Park and M. E. J. Newman, *The statistical mechanics of networks*, Phys. Rev. E **70**, 055101 (2004)

E. Ravasz, A. L. Somera, D. A. Mongru, Z. N. Oltvai and A. L. Barabási, *Hierarchical Organization of Modularity in Metabolic Networks*, Science **297**, 1551-1555 (2002)

S. S. Shen-Orr, R. Milo, S. Mangan and U. Alon, *Network motifs in the transcriptional network of Escherichia coli*, Nat. Genet. **31**, 64-68 (2002)

D. Sprinzak and M. B. Elowitz, *Reconstruction of genetic circuits*, Nature **438**, 443-448 (2005)

A. Wagner, *The yeast protein interaction network evolves rapidly and contains few redundant duplicate genes*, Mol. Biol. Evol. **18**, 1283-1292 (2001)

D. J. Watts and S. H. Strogatz, *Collective dynamics of 'small-world' networks*, Nature **393**, 440-442 (1998)

D. M. Wolf and A. P. Arkin, *Motifs, modules and games in bacteria*, Curr. Op. Microbiol. **6**, 125-134 (2003)

E. Yeger-Lotem, S. Sattath, N. Kashtan, S. Itzkovitz, R. Milo, R. Y. Pinter, U. Alon and H. Margalit, *Network motifs in integrated cellular networks of transcription-regulation and protein-protein interaction*, Proc. Natl. Acad.

Sci. USA **101**, 5934-5939 (2004)

Index

acetylation, 168
adiabatic, 5
Affymetrix chips, 117
alpha-helix, 95
amino acids, 88, 92
anti-sense strand, 84
applied work, 54
arc diagram, 89
asymmetric random walk, 52
autocorrelation function, 42
autoregulatory, 187
average degree, 230
average ignorance, 2

bending energy, 34
Bessel function, 170
beta-sheet, 95
bimodal distribution, 175
binding free energy, 157
biochemical pathways, 210
biochip, 117
biological network, 229
Bjerrum length, 159

canonical distribution, 4
canonical ensemble, 3
cDNA, 114
characteristic function, 12
chemical master equation, 66
chemical potential, 4, 8
chemical reactions, 65
chromatin, 167
chromatin remodelling, 168
circle diagram, 92
clustering coefficient, 233
coefficient of variation, 233
collision diagram, 65
communication channel, 218

comparative information, 145
compositionality, 219
conditional probability, 42
conservation laws, 66
contact matrix, 89
contour surfaces, 7
cooperative effect, 99
cooperative interaction, 176
cooperative phenomenon, 32
cooperativity parameter, 106, 108
copy number, 215
correlation length, 16
covariance, 12
covariance matrix, 12
critical exponent, 24
Crooks fluctuation theorem, 56
cross-hybridization, 117
cumulant expansion, 54
cumulants, 12
Curie's law, 27

Debye screening length, 160
Debye-Hückel equation, 160
denaturation bubbles, 100
denaturation loop, 100
dense overlapping regulon, 242
detailed balance, 51
diffusion equation, 43
diffusion matrix, 49
diffusion term, 57
directed polymer, 73
disorder, 176
dissipated work, 54
dissipation-fluctuation, 50, 61, 208
dissipation-fluctuation theorem, 41
division free energy, 141
DNA, 83
DNA microarray, 117

T - #0408 - 071024 - C8 - 234/156/12 - PB - 9780367390822 - Gloss Lamination